The
WATCH REPAIRER'S
MANUAL

The WATCH REPAIRER'S MANUAL

HENRY B. FRIED

Certified Master Watchmaker and Clockmaker
Technical Director
and President Emeritus
American Watchmakers Institute
Former Head, Department of Horology
George Westinghouse Vocational and Technical High School
New York City

Third Edition

CHILTON BOOK COMPANY
RADNOR, PENNSYLVANIA

Copyright © 1949, 1961, 1973 by Henry B. Fried
Third Edition   All Rights Reserved
Published in Radnor, Pa., by Chilton Book Company
and simultaneously in Ontario, Canada, by
Thomas Nelson & Sons, Ltd.
Manufactured in the United States of America

LIBRARY OF CONGRESS CATALOGING IN PUBLICATION DATA

Fried, Henry B   1907-
  The watch repairer's manual.

  Bibliography: p.
  1. Clocks and watches—Repairing and adjusting.
I. Title.
TS547.F7   1973   681'.114   73-10193
ISBN 0-8019-5916-0

PREFACE

The purpose of this book is to aid those interested in the practical art of horology. It is intended to serve as a course of study, as a means of reference, and as a textbook or working manual to be used in the school, at the bench, or in the library.

It is specifically aimed at serving the student receiving formal training in a horological school and to aid the teacher in that school. It should be equally helpful to the apprentice who is learning his trade "on the job," and to the hobbyist or collector who desires to learn the accepted methods of repairing watches.

The trade of watch repairing is made up of many individual jobs and operations, some elementary, others complex. The opening chapters are therefore introductory and simple in character. The chapters that follow deal with more difficult tasks, based upon work covered previously. Jobs that naturally follow previously completed operations are covered in the very next chapter. Drawings are used to illustrate manipulations not otherwise easily understood and to picture correct and accepted working methods.

Summary and test questions at the end of each chapter should permit this book to be used as a text and as part of the formal course of study in the classrooms of the horological school. Those who are using the book for self-study will find the summary and questions invaluable as a means of checking what they have learned.

The inclusion of a glossary of trade terms, tables, selected bibliography and other reference material should prove of especial help to the journeyman repairer, parts maker, watch material jobber, and others in the industry.

The success of the first edition has been manifested by the need for numerous printings. This new, third edition brings the instruction up-to-date, introducing new descriptions and new methods developed since previous editions were published. The inclusion of an additional chapter on inserting and fitting watch crystals should be most welcome and helpful to the reader.

March 1973   HENRY B. FRIED
Fresh Meadows, N.Y.

CONTENTS

The
WATCH REPAIRER'S
MANUAL

I

THE MODERN WATCH

U ntil recent times the invention of the watch was placed at the beginning of the sixteenth century in Nurnberg. Now, it has been established that portable timepieces existed in the latter part of the fifteenth century in Italy and northern France. Before that time, weights acting with gravitational pull served as the power unit. Naturally, this limited the clock to a position on a shelf, wall or the tower of some sturdy building. The creation of watches became possible when a strip of iron or steel could be fashioned to resemble a tape which could be wound around an axle fastened to the largest gear in the clock or watch. The force created by the unwinding of the springy metal tape rotated the wheels. Because this arrangement could be made small enough, it became possible to create a timepiece that could be carried on the person and eventually still smaller, allowing it to be carried in the pocket.

At first, all parts were made by hand, and, in time, many such small timepieces were made, each an improvement over previously made watches and therefore, at first, no two were alike. The development of the watch progressed through the centuries, each a hand-made product until the middle of the nineteenth century when American mass production methods introduced some degree of uniformity in manufacture and production. Today, a modern watch movement shows the results of almost five hundred years of steady development.

The Modern Watch Movement

The modern watch movement is a standard piece of machinery. Most movements have the same amount of wheels, springs and other parts. Fur-

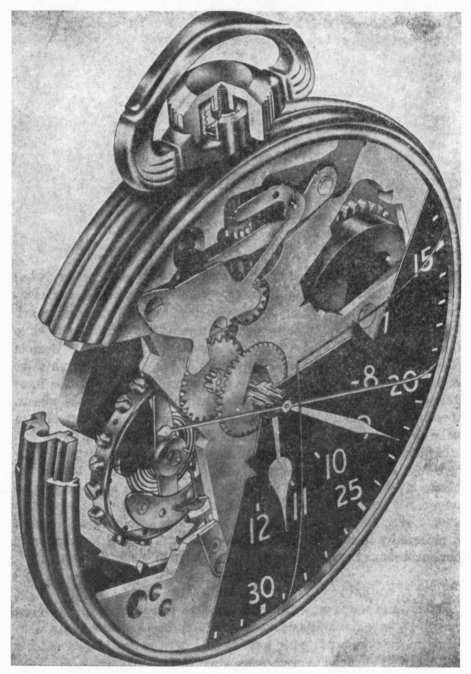

FIG. 1. Cutaway view of modern center-seconds pocket watch.

MOVEMENT AND CASE

1. Dial
2. Hour hand
3. Minute hand
4. (Sweep) Seconds hand
5. Case bow
6. { 7. Crown
   { 8. Stem
9. (Case) back cover
10. Case frame
11. Bezel
12. Movement
13. Watch crystal
14. Case screw

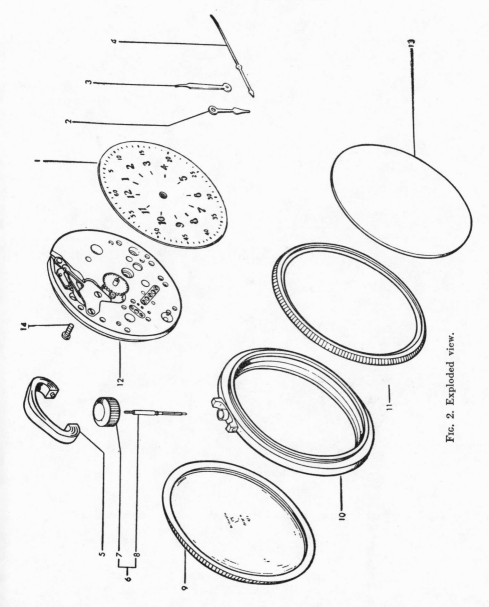

Fig. 2. Exploded view.

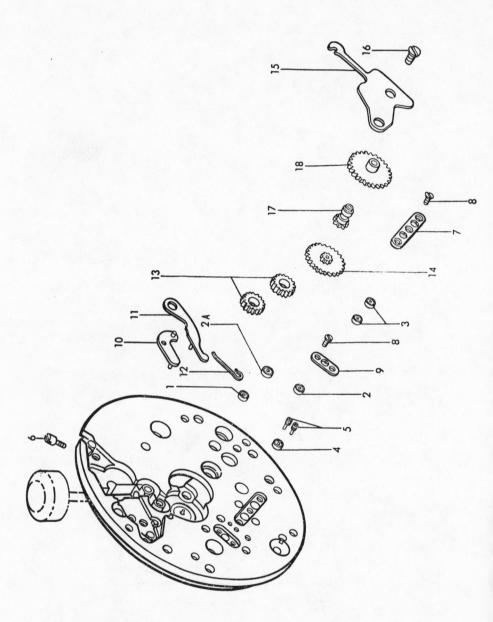

FIG. 3. Exploded view and nomenclature, dial side, of modern center-seconds watch.

thermore, they are arranged in a uniform pattern so that in most watches each part will be found in its corresponding place.

Each part performs its particular function in the watch. Therefore, it will bear characteristics determined solely by its purpose and differing only in size and minor details from a similar part in other watches.

The watch movement, in a sense, does not tell time but merely permits its mechanism to be released at a determined rate of speed. The dial and pointers (hands) indicating this speed translate it into terms measuring the passage of time. Thus, if this mechanism is released at a greater than normal speed, we say that our watch is going fast. If the watch movement should become sluggish and the gears and springs release themselves slowly, the hands indicate it as losing time.

The watch movement is made up of six main divisions:

1. The power unit
2. The winding mechanism
3. The train of wheels
4. The escapement
5. The dial train
6. The setting mechanism.

Like any other piece of machinery, the watch movement must have power. Because of the compactness of this small mechanism, the power unit also must be small, compact and housed within the watch. The power unit (Figure 5) is composed of a mainspring, a thin, narrow ribbon-like piece of fine spring steel attached at its inner end to a hook on a miniature spool called the barrel arbor. This barrel arbor serves as the axle of the barrel.

The barrel is a round, flat, hollow box with gear teeth around its outer circumference. The mainspring with its inner end attached to the barrel arbor is wound around the arbor and inserted into the barrel.

DIAL SIDE OF MOVEMENT (Exploded View)

1. Third Wheel Lower Hole Jewel.
2. Fourth Wheel Lower Hole Jewel.
2A. Center Wheel Lower Hole Jewel.
3. Lower Plated Pallet and Escape Wheel Hole Jewels.
4. Lower Balance Hole Jewel.
5. Pallet Banking Pins (Set Eccentrically in Screws).
6. Dial Screws.
7. Lower Pallet and Escape Wheel Cap Jewel.
8. Cap Jewel Screws.
9. Lower Balance Cap Jewel.
10. Set Lever (Detent).
11. Clutch Lever.
12. Clutch Lever Spring.
13. Intermediate Setting Wheels.
14. Minute Wheel.
15. Setting Bridge.
16. Setting Bridge Screws.
17. Cannon Pinion.
18. Hour Wheel.

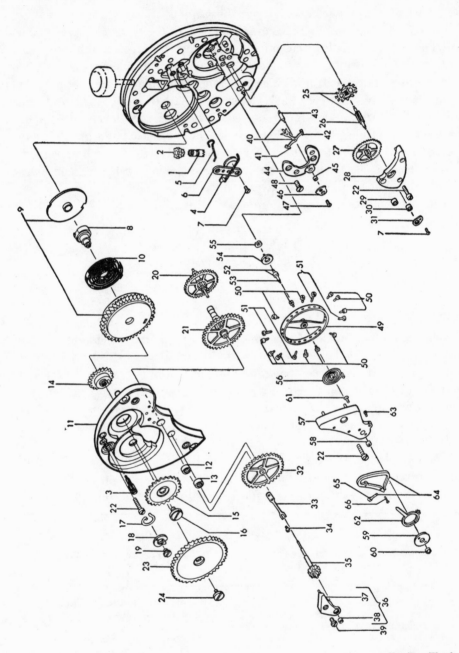

FIG. 4. Exploded view and nomenclature, movement side, of modern center-seconds pocket watch.

The outer end of the mainspring is attached to a brace or connection to the inside wall of the barrel. If the barrel is held in the fingers of the left hand and the arbor is turned with the other hand, the mainspring would be wound around the arbor. This spring, in its efforts to resume its previous state of rest, would exert a force upon its outer end, which is attached to the barrel, and would pull the barrel around with it in its efforts to uncoil itself.

The gear teeth around the barrel's edge will propel any pinion or gear enmeshed with them as the mainspring forces the barrel around. (See Figure 6.)

The Winding Mechanism

To store up the power inside the barrel, a ratchet wheel is attached to the barrel arbor. When the ratchet wheel is rotated, it turns the arbor and

---

TRAIN SIDE OF MOVEMENT (Exploded View)

1. Clutch Wheel.
2. Winding Pinion.
3. Detent (Set Lever) Screw.
4. 5. 6. 7. } Balance Stopping Mechanism, used mainly for Navigational and Military Use. (Hackwatch)
8. Barrel Arbor.
9. Barrel and Cover.
10. Mainspring.
11. Barrel Bridge.
12-13. Center and Third Wheel Upper Hole Jewels in Settings.
14. Lower Crown Wheel. 15. Upper Crown Wheel. } Winding Wheel
16. Crown Wheel Screw.
17. Clickspring.
18. Click.
19. Click Screw.
20. Third Wheel.
21. Center Wheel.
22. Bridge Screw.
23. Ratchet Wheel.
24. Ratchet Wheel Screw.
25. Escape Wheel.
26. Escape Wheel Pinion.
27. Fourth Wheel.
28. Train Bridge.
29-30. Train Wheel Hole Jewels In Settings.
31. Escape Wheel Cap Jewel.
32. Sweep Second Wheel.
33-34. Tension Spring and Screw.

35. Center Seconds Pinion.
36. { 37. Center Seconds Pinion Bridge. 38. Center Seconds Bridge Hole Jewel. 39. Center Seconds Bridge Screw.
40. Pallet.
41. Entrance Pallet Jewel.
42. Exit Pallet Jewel.
43. Pallet Arbor.
44. Pallet Bridge.
45. Pallet Bridge Hole Jewel.
46. Pallet Bridge Cap Jewel.
47. Cap Jewel Screw.
48. Pallet Bridge Screw.
49. Balance Wheel.
50. Balance Screws.
51. Timing Screws (Adjustable).
52. Balance Staff.
53. Balance Staff Pivot.
54. Impulse Roller and Jewel.
55. Safety Roller.
56. Hairspring.
57. Balance Bridge.
58. Balance Bridge Hole Jewel (Set).
59. Balance Bridge Cap Jewel Plate.
60. Cap Jewel (Bal. Bridge) Set.
61. Balance Bridge Cap Jewel Screw.
62. Regulator. (Breguet)
63. Stud Screw.
64. Regulator Spring.
65. Regulator Spring Micrometer Screw.
66. Regulator Spring Screw.

winds up the attached spring. Because the mainspring exerts a force against the ratchet wheel, a retaining pawl, or *click*, is made to engage the ratchet wheel teeth so that the ratchet wheel can move only in the winding direction. The ratchet wheel is turned by a series of gears connected to the winding button or *crown*. This series of gears is called the winding mechanism (Figures 6 and 7).

The winding mechanism consists of a crown attached to a stem. The

Rotherham and Sons Ltd.

Fig. 5. The power unit of the watch is the mainspring
wound within the barrel.

square length of this stem is inserted into the square hole of the clutch wheel so that, when the stem and crown are turned, the wheel also turns.

Above the clutch wheel is the winding pinion. It has a round hole and cannot be turned directly by the stem. The winding pinion has two sets of teeth set like a crown. One group is radial, pointing outward around its circumference, whereas the other group of teeth is planted on a smaller diameter, perpendicular to the first set. The perpendicular set of teeth are ratchet or slanted teeth. These are made to mesh with the clutch teeth so that the winding pinion moves only when it is enmeshed with the clutch wheel and only in one direction.

When the winding pinion is turned by the clutch wheel, the winding pinion's radial teeth meshes with another wheel at right angles to it and directly behind the stem in the back of the movement. This wheel is called the crown or *winding wheel*. In some watches it may have two sets of teeth —one engaged by the winding pinion, and one set engaging the ratchet wheel mounted on the mainspring barrel arbor. This action winds up the watch.

The Train

The mainspring and barrel drive the train. The train consists of four wheels: the center wheel, the third wheel and pinion, the fourth wheel and pinion, and the escape wheel and pinion.

The purpose of the train is to supply energy to the escapement in small units. If the mainspring was connected directly to the escapement, the power supplied would be comparatively tremendous. Also, the power would be quickly spent and the mainspring would then require frequent rewinding.

The train actually is a set of reduction gears designed to diminish the mainspring's power and expend it a little at a time so that it will last longer. The center wheel generally makes the third wheel turn 8 times

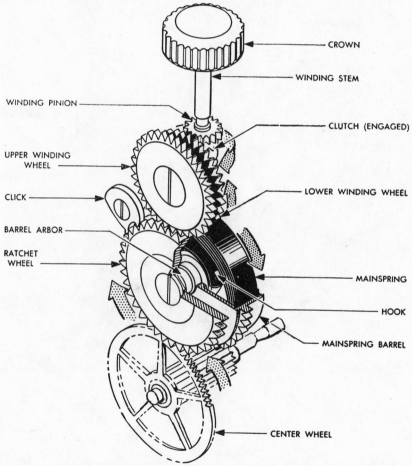

FIG. 6. The winding mechanism is used to store the power in the mainspring.

while it turns once. The third wheel, in turn, rotates the fourth wheel pinion and wheel, multiplying it about $7\frac{1}{2}$ times or at a ratio of 1:60 with the center wheel. The fourth wheel turns the escape pinion and wheel 10 times to each of its own revolutions. The escape wheel contains 15 teeth, each of which work on two levers (Pallet jewels) so that the multiplication totals 18,000 vibrations of the balance to one turn of the center wheel. The center wheel is turned about 8 times for each coil of wound mainspring. About 5 full windings of this spring keep the watch going 40 hours.

The center wheel, largest in the train, gets its name because of its position in the watch. This wheel is mounted on the center pinion or arbor. Also called the center post, this pinion is the longest in the watch because

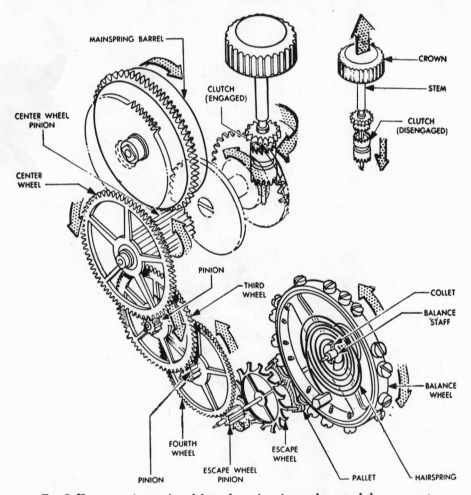

FIG. 7. How power is transferred from the mainspring to the rest of the movement.

upon it ride the hands. This post connects the train with the dial-indicating mechanism (dial train).

The barrel drives the center pinion to which is attached the center wheel. The center wheel drives the third pinion and, with it, the third wheel, which is part of the train. The third wheel also acts as an intermediate wheel between the hand-carrying center wheel and the second-hand carrier, the fourth wheel. If the fourth wheel were to be driven directly by the center wheel, the second hand would be turning counterclockwise. The intervening third wheel, therefore, in addition to helping in the reduction of the mainspring's power into small units, also keeps the second hand turning in a clockwise direction.

Rotherham and Sons Ltd.

FIG. 8. The train.

The fourth wheel generally turns 60 times the number of revolutions of the center wheel. It is engineered so that it will carry a second hand on its long tapered pivot. The fourth wheel is placed at right angles to the stem or on a line diametrically opposite the stem, depending, of course, upon the design of the dial.

The escape wheel pinion is the last part in the train, but its wheel belongs to the escapement.

The Escapement

The escapement (Figures 9 and 10) consists of three main parts—the escape wheel, the pallet, and the balance. The escapement is the heart and brain of the watch. Upon its precision, quality, and condition depends the purpose of the watch—namely, to keep accurate time.

The purpose of the escapement is to regulate the flow of power through

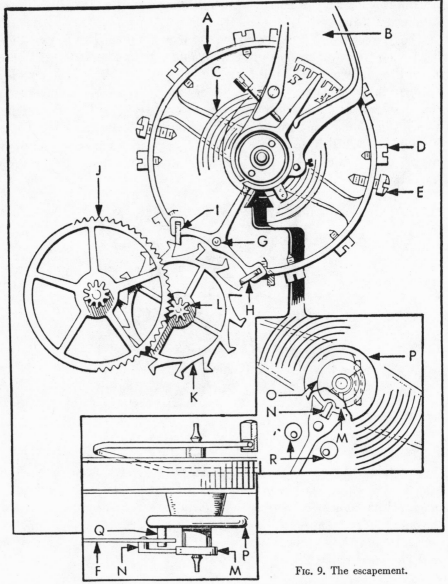

Fig. 9. The escapement.

A. Balance Wheel
B. Balance Cock
C. Hairspring
D. Weight Screw
E. Meantime Screw
F. Pallet
G. Pallet Arbor
H. Let-off Pallet Jewel
I. Receiving Pallet Jewel

J. Fourth Wheel
K. Escape Wheel
L. Escape Wheel Pinion
M. Small Roller
N. Pallet Guard Pin
O. Hairspring Collet
P. Large Roller
Q. Roller Jewel
R. Banking Screws (2)

the wheels so that these turn at a steady rate of speed coincident with the passing of time. The hands indicate this speed.

Rotherham and Sons Ltd.

FIG. 9a. The pallet and escape wheel and pinion.

Almost all escape wheels have 15 teeth, each shaped like a flat-soled boot. These wheels, unlike the brass train wheels, are made of thin, hard polished steel with beveled tooth ends. This wheel generally makes 10 revolutions to each of the fourth wheel. It takes 6 seconds to make one full revolution.

The pallet is in the form of a T with an oblong slant-edged jewel planted on each end of the T bar. The pallet is pivoted at the cross of the T. The escape wheel with its flat-edged teeth act upon the slanted, polished end surfaces of the pallet jewels and lift or push up the pallet so that the forked tail moves from one side to the other.

In this way a tooth of the escape wheel passes or escapes, permitting the entire train to advance that little

Rotherham and Sons Ltd.

FIG. 10. The balance, complete.

bit while another tooth locks near the edge of the pallet jewel.

Resting in the forked tail of the pallet is a D-shaped pin attached to the balance wheel. When the pallet is moved, it will push the jewel pin. This will cause the balance to be moved away.

The hairspring which is attached to the other side of the balance and anchored to the movement is wound up by the turning motion of the balance. The hairspring will then force the balance back again and, upon its return, it will have enough momentum to move the fork and unlock the escape tooth now upon the opposite pallet jewel. This tooth will then push the pallet back and this, in turn, will give another impulse to the balance in the direction of its own momentum. The balance will continue to move until arrested by the resistance of the hairspring. This process will repeat itself as long as there is power to drive the escape wheel.

The speed with which the balance moves back and forth (which directly influences the speed of the watch) is governed by the strength of the hairspring. The hairspring is truly named, for it is a flat, spiraled spring no thicker than a human hair. It is connected to a brass collar that is held friction tight upon the axle of the balance.

The balance is the most complicated and expensive unit within the watch movement. Its rim is composed of metals or alloys that prevent its undue expansion or contraction during temperature changes. Along this rim are threaded a number of screws set in balanced and matched pairs, each diametrically opposite the other. These are used to add to its mass and permit adjusting. The balance is poised so that no one screw exerts a greater gravitational pull than any other.

The balance rotates on an axle called the *balance staff*. This staff has two hair-thin pivots, highly polished and revolving in jeweled bearings. The pivot-ends rest on solid, flat jewels.

On the lower part of the balance staff are mounted two small discs, one about half the size of the other. These together form the *roller table*. Into the larger one, a D-shaped jewel pin called the *roller jewel* is set upright. This is the jewel pushed by the pallet fork. The lower, smaller disc has a crescented slot and works in conjunction with the little pin sticking out of the fork slot. It is called the *guard finger* and is used to keep the pallet from becoming unlocked prematurely.

In most watches, the balance and hairspring are co-ordinated to swing back and forth *(vibrate)* 18,000 times an hour, or 300 times a minute, or 5 times a second. As the fork finishes its thrust upon the balance, it locks an oncoming escape tooth. Thus, while the balance is moving around, the watch is at rest until the balance unlocks the escapement and the escape tooth drops on the other pallet jewel. As the escape wheel moves, the second hand on the fourth wheel behind it moves forward one-fifth of a second. It will take therefore five of these "escapes" for the second hand to advance one full second on its dial.

Bearings

In most watches, the train and escapement have pivoted arbors and run in jeweled bearings. The pivots are thin, hard and highly polished to resist friction and wear. The jeweled bearing holes are also highly polished, and they aid in the longevity of the timepiece. Jewels used for bearings are garnet, sapphire, and ruby. In most modern watches, the better jewels are made from synthetic ruby, garnet, or sapphire because these are purer and contain less flaws for their industrial use in the watch movement.

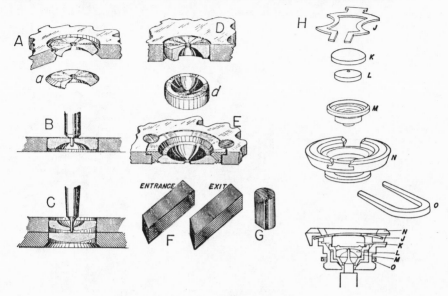

Fig. 11. Jewels used in watches: A is an old-fashioned bevel-edged hole jewel, held in plate by burnished over *bezel* setting, no longer used in modern watches. *a* is bevel-edged jewel unset. B is a friction (press-fit) jewel as used today. C is a conical pivot resting on a cap jewel (endstone) and contained by the olived-holed balance jewel. Both are of the modern friction type. D is a friction plate jewel press-fitted into a plate. *d* is the same jewel unset. E shows a bezel set jewel contained in a separate bushing held in the plate by screws. F are pallet jewels, the entrance and exit. Notice their different face angles. G is a roller jewel. H shows an exploded view of a popular type of shock proof jewel assembly (Incabloc). J is the shock spring. K the cap jewel, L the balance hole jewel, M the bushing for the hole jewel and seating for the cap jewel. N is the main bushing receptacle. O is the locking clamp. The lower figure shows the unit assembled.

The Dial Train

To indicate the time in hours and minutes, a special set of gears is situated underneath the dial. These serve as miniature reduction gears, so that the hourly revolution of the center pinion and minute hand (Figure 12) may by its transmission turn the hour hand once in 12 hours. The dial train is composed of three main parts: the cannon pinion, the minute wheel and pinion, and the hour wheel.

The cannon pinion is a small, steel tube with teeth resembling daisy petals planted on its lower edge. This pinion slips with a clutch-like grip over the part of the center post emerging through the dial side of the movement. The teeth or *leaves* of the cannon pinion mesh with the teeth of the minute wheel.

The minute wheel, set beside the cannon pinion under the dial, has a flat, small pinion. The minute wheel idles upon a pin set in the movement and sticking up through the hole in the minute-wheel pinion. The hour wheel rests loosely over the cannon pinion. It is usually a thin, brass wheel with

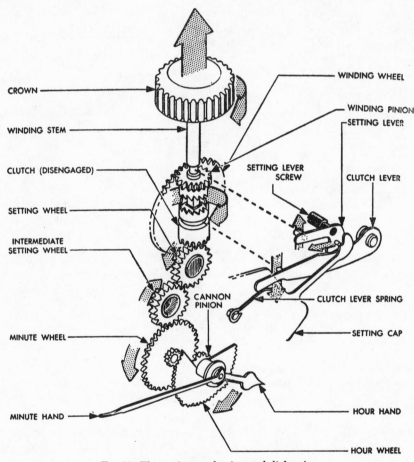

CROWN

WINDING STEM

CLUTCH (DISENGAGED)

SETTING WHEEL

INTERMEDIATE
SETTING WHEEL

MINUTE WHEEL

MINUTE HAND

WINDING WHEEL

WINDING PINION
SETTING LEVER

SETTING LEVER
SCREW

CLUTCH LEVER

CANNON
PINION

CLUTCH LEVER SPRING

SETTING CAP

HOUR HAND

HOUR WHEEL

Fɪɢ. 12. The setting mechanism and dial train.

a thin, hollow tube. The top of this tube is lower than that of the cannon pinion. The hour wheel meshes with the minute-wheel pinion. Whenever the center pinion post turns, it carries along with it the cannon pinion.

The cannon pinion, let us say, has 10 teeth. The cannon pinion turns the minute wheel which has 30 teeth. Therefore, to turn the minute wheel around once, the cannon pinion must make 3 revolutions. The minute-wheel pinion attached to its wheel also has 10 teeth. This pinion meshes with the hour wheel loosely atop the cannon pinion. The hour wheel has 40 teeth. In order for the hour wheel to turn around once, the minute-wheel pinion, together with the minute wheel, must make 4 complete revolutions. For each turn of the minute wheel the cannon pinion must revolve 3 times. Thus, for every 4 turns of the minute wheel, the cannon pinion must make 12 turns and, in doing so, the hour wheel will turn only once. The hour hand is placed on the hour wheel, and the minute hand is placed on the cannon pinion.

The Setting Mechanism

The setting mechanism is the connecting link between the winding mechanism and the dial train. Most of the time, the dial train is motivated by the revolving center pinion. However, some control is needed to adjust the hands to indicate correct time. The setting mechanism is composed of a series of levers that shift the movement of the stem and clutch wheel from the winding wheels to engage the dial train wheel. (See Figure 12.)

The setting mechanism consists of a setting lever or detent; a clutch lever, clutch lever spring, setting cap or setting bridge, and one or two setting wheels also called intermediate (setting) wheels. The setting lever pivots on a free-riding shoulder screw called the setting lever screw. The forward end of the setting lever has a pin that fits into the neck of the stem. When the crown and stem is pulled up, the setting lever pin moves up with it and the rear end see-saws downward. The clutch lever is pushed downward by the rear end of the setting lever. The finger-like clutch lever, resting in the neck of the clutch wheel, moves this wheel downward and disengages it from the winding pinion.

The lower end of the clutch wheel, with its teeth pointing downward, engages the setting wheels. These are idler wheels connected to the dial train. A slight groove in the clutch lever keeps the setting lever in place. To aid in keeping the setting lever in the setting position, the setting cap has a finger-like spring which locks the thin pin set upright through the rear of the setting lever. The setting cap also keeps in place the clutch lever, minute wheel, and intermediate wheel, as well as the clutch lever spring.

The crown and stem, turning in the direction of the arrow (Figure 12), rotate the clutch wheel and dial train wheels and hands as shown in this illustration.

If the crown is pushed downward, the clutch lever is released and the clutch lever spring will push the clutch wheel back into engagement with the winding mechanism.

Plates and Bridges

All the wheels and pinions, springs and screws, are contained within the bridges and plates comprising the movement. The movement consists of a main plate which serves as the skeleton upon which the rest of the movement is built. The bridges cover the wheels and serve to bear their pivots.

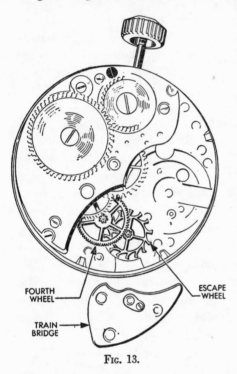

Pivots of all wheels resting in bearings in the main plate are called *lower* pivots. To support and house the bearings for the upper pivots of these wheels, various bridges are used.

The barrel bridge houses the barrel and arbor, ratchet wheel, click, and the crown (winding) wheel. Sometimes this bridge also contains the center wheel or additional train wheels.

FOURTH WHEEL

ESCAPE WHEEL

TRAIN BRIDGE

FIG. 13.

The bridge that supports the upper pivots of the train wheels is called the *train bridge*. Some watches have individual bridges for each wheel.

Wheels differ from pinions in shape and purpose. Wheel teeth are shaped with straight sides and have bullet-shaped ends. Pinion leaves have radial sides and rounded ends. These are generally shaped like daisy petals. Wheel teeth are the drivers; they have received the power from the mainspring and do the pushing. Pinion leaves are the followers; they are pushed by the wheel teeth.

Wheel teeth utilize only their curved tips to move the pinions. The pinion is pushed only on its straight radial flank. The rounded tops are provided for streamlining and as a safety measure. The spaces between wheel teeth provide clearance for the pinion leaves. (See Figure 14.)

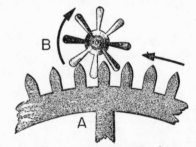

FIG. 14. The wheel *A* is the driver, pushing pinion *B*.

The above is true of all wheels in the watch with the exception of the dial train. Here the cannon pinion and minute-wheel pinion push the other dial train wheels while the watch is going. When the crown and stem take over the setting of the hands, the wheels push the pinions. Therefore, the dial-train pinions are shaped in a sort of compromise design.

Summary

1. The movement consists of six main divisions: the power unit, the winding mechanism, the train of wheels, the escapement, the dial train, and the setting mechanism.

2. The power is supplied in the form of a coiled spring housed in a geared barrel.

3. The winding mechanism is a series of gears which connects the winding stem to the mainspring, permitting the spring to be wound.

4. The "train" is composed of four wheels geared to reduce the power of the mainspring into smaller units. These wheels are: the center wheel, the third wheel, the fourth wheel, and the escape wheel. The train wheels carry the power from the mainspring to the escapement.

5. The escapement is composed of the escape wheel, pallet, and the balance. Its purpose is to permit the orderly progress of the train wheels and hands. It is the unit in the movement responsible for the timekeeping of the watch.

6. The dial train is composed of three parts: the cannon pinion, minute wheel and pinion and the hour wheel. These are situated under the dial and drive the hour and minute hands.

7. The setting mechanism is a set of levers and springs that shift the clutch wheel from the winding position to enmesh with the dial train and make it possible to set the hands through the crown and stem.

8. The plates and bridges house the movements and act as the skeleton upon which the rest of the movement is built.

9. Wheels are pushers and pinions are followers pushed by the wheel teeth.

10. Most watch movements "vibrate" 18,000 times an hour.

Questions

1. Name the main divisions of a modern watch movement.
2. Name the component parts of each division.
3. Describe the purpose of each unit.

4. Upon what part of the watch is placed the minute hand? The hour hand? The second hand?

5. What is the purpose of the click?

6. How do wheel teeth differ from pinion leaves? What difference is there in purpose?

7. What is the ratio between the cannon pinion and the hour wheel?

8. What is the clutch wheel? The intermediate wheel?

9. In what direction does the clutch wheel grasp the winding pinion? Why?

10. For what purpose is the setting cap, or setting bridge, used?

11. For what purpose is the balance wheel used?

12. In what unit of the watch will the pallet be found?

13. When was the first watch invented?

14. Where is the jewel pin situated?

15. What type of bearings support the pivots of the balance staff?

II

CLEANING AND OVERHAULING A WATCH MOVEMENT

Like any other machine, a watch movement needs periodic overhauling and lubrication. Cleaning a watch movement generally includes washing the parts to remove any foreign matter, such as lint and dust. These small particles will stop the watch because they will clog the tiny teeth and pinion leaves.

Oil that has dried and become gummy acts as a further hindrance, so that the mainspring's power can no longer overcome the resistance offered by the combination of dirt and gummed oil. When this happens, the watch must be dismantled and each part washed, dried, reassembled, and lubricated. Wrist watches should be overhauled between 9 and 12 months after their last servicing. Pocket watches should be serviced at least once a year.

Many watch parts are made of numerous smaller pieces. Sometimes it is not necessary to dismantle such a unit. For instance, the complete balance wheel consists of a hairspring, rollers, and an average of 14 screws. This unit is not dismantled but treated as one piece.

To illustrate the cleaning process, a wrist-watch movement shown in the exploded view (Figure 1) will be the model used. Wrist-watch movements are usually contained in two piece cases—a bottom which houses the movement, and the top which snaps onto the bottom. The top contains the crystal and provides a fastening for the bracelet or strap.

The watch case is opened by placing the blade of a knife or "case opener" on the lip extending from the bottom of the case and by using a prying motion to lift the case top. The movement is removed from the bottom of the case by placing the thumbnail on the ledge of the case bottom close to the crown and the index finger under the crown and pushing upwards. The index finger and thumb of the left hand grip the ends of the

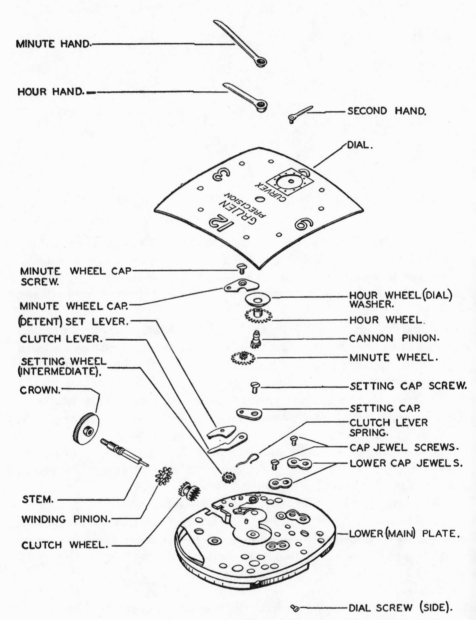

MINUTE HAND.

HOUR HAND.

SECOND HAND.

DIAL.

GRUEN PRECISION CURVEX
12

MINUTE WHEEL CAP SCREW.

MINUTE WHEEL CAP.

(DETENT) SET LEVER.

CLUTCH LEVER.

SETTING WHEEL (INTERMEDIATE).

CROWN.

HOUR WHEEL (DIAL) WASHER.

HOUR WHEEL.

CANNON PINION.

MINUTE WHEEL.

SETTING CAP SCREW.

SETTING CAP.

CLUTCH LEVER SPRING.

CAP JEWEL SCREWS.

LOWER CAP JEWELS.

STEM.

WINDING PINION.

CLUTCH WHEEL.

LOWER (MAIN) PLATE.

DIAL SCREW (SIDE).

FIG. 1a. Exploded view and nomenclature, dial side of modern wrist watch.

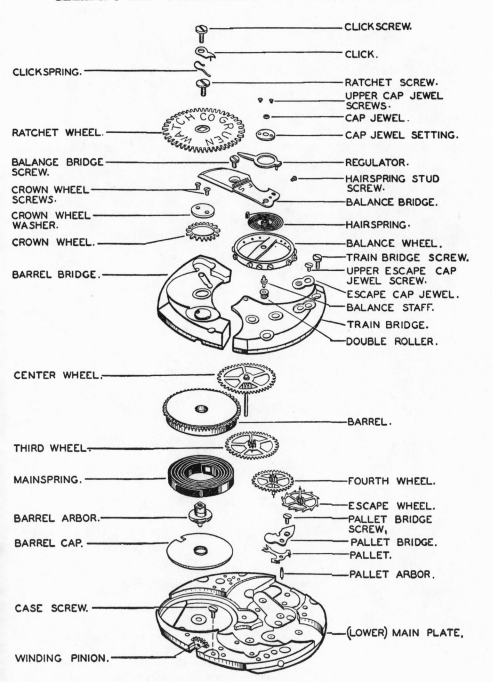

CLICK SCREW.

CLICK.

CLICK SPRING.

RATCHET SCREW.

UPPER CAP JEWEL SCREWS.

CAP JEWEL.

RATCHET WHEEL.

CAP JEWEL SETTING.

BALANCE BRIDGE SCREW.

REGULATOR.

HAIRSPRING STUD SCREW.

CROWN WHEEL SCREWS.

BALANCE BRIDGE.

CROWN WHEEL WASHER.

HAIRSPRING.

CROWN WHEEL.

BALANCE WHEEL.

TRAIN BRIDGE SCREW.

BARREL BRIDGE.

UPPER ESCAPE CAP JEWEL SCREW.

ESCAPE CAP JEWEL.

BALANCE STAFF.

TRAIN BRIDGE.

DOUBLE ROLLER.

CENTER WHEEL.

BARREL.

THIRD WHEEL.

MAINSPRING.

FOURTH WHEEL.

ESCAPE WHEEL.

BARREL ARBOR.

PALLET BRIDGE SCREW.

BARREL CAP.

PALLET BRIDGE.

PALLET.

PALLET ARBOR.

CASE SCREW.

(LOWER) MAIN PLATE.

WINDING PINION.

FIG. 1b. Exploded view of movement side of same watch.

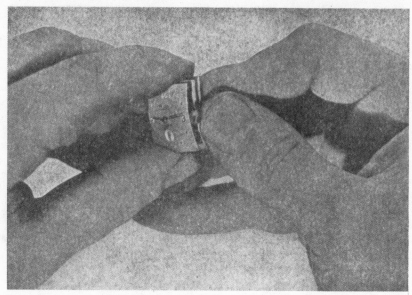

Photo by Gruen Watch Co.

FIG. 2. Easing the movement out of the case.

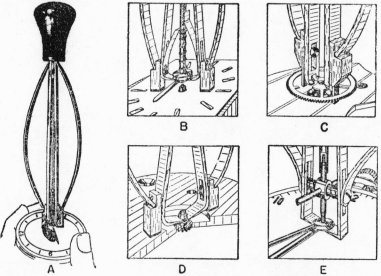

FIG. 2A. A "puller" type of hand remover and its other uses with different prongs. A and B, removing hands. C, removing sweep second wheels. D, removing cannon pinions, E, remover with center post removing large watch hands.

dial and ease the movement out of the case. This operation must be carefully performed and this is shown in Figure 2. The use of a knife to pry out the movement may cause damage.

The hands of the watch may be removed while the movement is in the case bottom or after it is taken out. Figure 2-3 shows how these hands are removed. To protect the dial from being scratched, a dial protector is

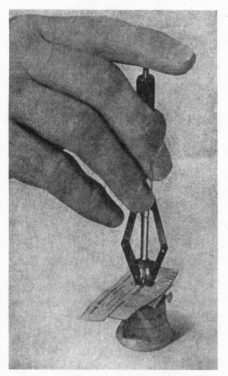

FIG. 3. The dial is protected by a thin celluloid strip while the hands are being removed.

FIG. 4. Another method of removing hands.

used. This is a paper-thin strip of celluloid with a V cut in the middle so that only the hands are exposed. In this way any pressure or scraping is received by the celluloid strip.

If the hands are removed while out of the case, the movement should be placed on a movement rest as shown in Figure 3. The dial is then removed (Figure 4) by loosening the side dial screws and lifting the dial, dial washer, and hour wheel (Figure 5).

Remove the balance and bridge assembly from the movement (Figure 6). This must be done carefully so that the balance or hairspring does not

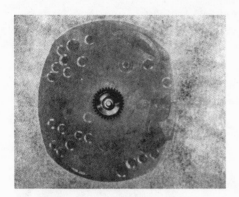

Fig. 5.   Fig. 6.

get fouled with the center wheel. Sometimes, the pivot seems to bind in the jewel. This difficulty may be due to thickened oil that has hardened around the pivot. In such an event, hold the balance bridge slightly over the movement while it is connected to the balance by the hairspring. Then, with the left hand, lift the movement about $\frac{1}{16}''$ above the movement rest and let it drop onto it. This slight jarring will loosen the pivot from the gummed jewel and the assembly may then be removed.

Separate the balance from the bridge. The stud screw is loosened before the balance bridge is removed because it is easier to do this while the bridge is attached to the movement.

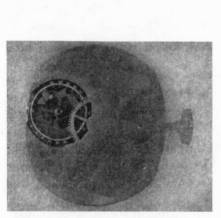

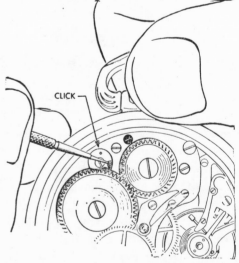

Fig. 7.

Fig. 8. The power stored in the mainspring must be released gently.

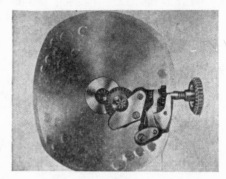

FIG. 9.   FIG. 10.

If the hairspring is of the overcoil type, the balance might separate from the bridge as it is being lifted out of the movement as shown in Figure 7.

Before proceeding with the dismantling, the power in the mainspring is released by placing a pointed piece of pegwood or a needle set into a piece of pegwood into the click. The click is moved out of the ratchet teeth and the mainspring is released while the crown is held in the fingers so that the power is not released too suddenly. A method of releasing the power is shown in Figure 8.

The pallet bridge and pallet are then removed (Figure 9). This is done by turning the movement over so that the dial side is up. The pallet-bridge steady pins may then be pushed out, causing the pallet and bridge to drop out of the movement. This method will prevent the bridge from becoming bent or scratched.

FIG. 11.   FIG. 12.

Next remove the cannon pinion and the dial train wheels and setting parts (Figure 10). The cannon pinion may be removed by using a pin-vise to grip and twist it off the center post.

Remove the ratchet and crown wheels (Figures 11 and 12). Remove the barrel bridge first and then the train bridge. When they have been removed, the train (Figure 13) will be exposed. In removing the train wheels, lift the wheel that is highest, then the next lower wheel until all the wheels are removed. As shown in Figure 13, in this case, the center wheel is removed first, the barrel is next, followed by the third, fourth, and escape wheels, in that order.

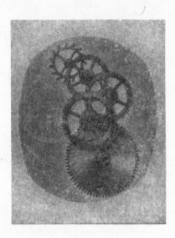

FIG. 13.

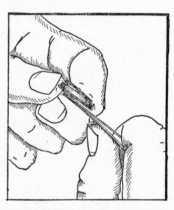

FIG. 14. A method of removing the barrel cover using the flat back of the tweezer.

Remove the barrel cap by prying it off through the small opening near its edge or by pushing up the barrel arbor from its screwhole side while the barrel is held by its edges between the fingers as shown in Figure 14. Remove the barrel arbor by carefully applying to it a clockwise and upwards pressure. The mainspring is also removed. This is done by grasping the inner coils with a strong pair of tweezers and permitting one half coil out at a time, releasing each coil by slackening the pinching pressure of the thumb and index finger of each hand alternately until the complete spring is removed.

Remove all cap jewels. All the parts except the dial and hands and case should be placed in a fine wire mesh cup similar to a tea strainer. The balance wheel is placed in last on top of the other parts.

The mesh basket is put into a cup about 3″ in diameter and about 2½″ deep that is nearly full of cleaning solution. Cleaning solutions may be purchased from jewelers' supply houses. Let the basket and parts soak in the solution for a few minutes, the specific length of time depending on the strength of the cleaning fluid. If the plates and wheels are badly tarnished, the parts may be brushed individually with the cleaning solution and a soft hair brush.

The mesh basket is then placed in another jar filled with benzene. The basket is shaken lightly to rinse off any remaining cleansing solution. This jar will then contain a mixture of soapy cleaning solution and benzene; therefore, another jar of clear benzene should be used for a final rinse.

Remove the balance from the first jar of benzene and rinse in the second jar of clear benzene. Dry the balance by blotting it gently with a Sylvet or clean linen cloth that will absorb the liquid clinging to the balance and hairspring. The drying must be done carefully so that the coils of the hairspring will not become distorted. The cloth used must be absolutely clean. Another method of drying the balance is to drop it into a box of warm boxwood sawdust. However, although the sawdust will absorb the moisture, tiny bits may cling to the balance and cause the watch to stop when they later become dislodged.

Blowing the parts dry while they are held in a piece of tissue paper by directing a jet of air from the bench blower is still another method used. There are other methods by which the watch parts can be dried. Using heat to dry the watch parts is often a risky operation both from the point of personal safety and damage to the parts. Most dehydrators recommended for horological use are volatile and inflammable. Overheating parts that have been soaked in these dehydrators may cause explosions or fires. Excessive heat, as well as disturbing the temper of the steel parts, may also melt the shellac that holds the pallet or roller jewels.

Take the remaining parts out of the rinse, one at a time, and dry them, after which they should be placed under a clear glass bell jar to protect them from dust or lint. A piece of pegwood, sharpened to a point, is used to clean out the confined area in the jewel holes and bearings. The pegwood must be sharpened with a knife so that its sides present faceted edges, similar to a broach. In this way, the sharp edges may remove any gummed oil or dirt clinging to the insides of the holes. The pegwood is applied from both sides of the jewel and is twisted into the hole to insure thoroughness.

The pegwood should be resharpened with the edge of a razor or scraped with a bit of broken watch crystal to remove any dirt that may cling to it after its removal from a jewel hole. This precaution is necessary so that the pegwood will not transfer dirt from one hole to the other. Figure 15 shows the method of applying the pegwood to the hole in a plate.

The best pegwood for horological use is made of boxwood. This type is about ⅛″ thick and about 6″ long. It will hold a fine point, has good absorbent qualities, will not shed, and is tough enough for practically all work.

All cap jewels must be wiped with a chamois buff or rubbed with the tip of a clean piece of pegwood until the jewel surface glistens. This is necessary to remove any haze or chemical film still adhering to the surfaces.

Pivots of all wheels should be cleaned by sticking them into soft pithwood. The spongy pith resembles cork in consistency but is much lighter. The pith does to the pivots what the chamois buff does to the cap jewels— removes any film and restores the fine black finish to the polished pivots.

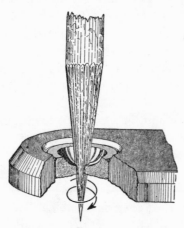

Fig. 15. How the pegwood is applied to a hole jewel to remove gummed oil or dirt.

Fig. 16. A telescopic type of oilcup. The upper reservoir is used for fine oil and the lower holds the heavier oil.

After pegging the holes and rubbing the pivots into the pithwood, all parts should be brushed lightly with a soft hair brush to remove any lint shed by the pithwood. The brush should be clean and used sparingly to avoid dust being placed again on the parts. All the parts should be inspected under the loupe to discover whether any dust, lint, or rust spots, bent pivots or other imperfections exist.

Assembling and Oiling

There are many commercial oils advertised, some of which have proved satisfactory for general repair work. Generally, there are three types of lubricants used in watchwork. Fine watch oil is used for the pivots of the train and escapement pivots. Clock oil for the mainspring and sometimes for the winding wheels. White vaseline jelly, however, is sometimes recommended instead of clock oil for the winding mechanism because of the heavier wear of these parts.

The watch movement is oiled at various stages of assembly. Oil that is old or that has been exposed to air becomes rancid and not worthy of use. The oil should be kept in a small glass or agate "well" (Figure 16). Called "oil cups," they are small discs of glass or agate about $\frac{1}{2}''$ in diameter set into round, wooden platforms. The center of the agate is ground to a concave shape about $\frac{1}{4}''$ in diameter or less. This countersink forms the *well*. The oil in the well should be changed often and the well cleaned before being refilled. The oil cups should be covered with a clean cover when not in use.

Oil applicators come in various sizes. Oilers are bits of hair-thin wire of different thicknesses whose ends are flattened slightly so that, when they are dipped into the oil cup, a tiny bead of oil will cling to the flattened end. When the oiled tip is applied to the pivot in its jeweled bearing, the oil is transferred to the pivot and bearing hole. Oilers are usually set into small wooden or plastic handles. Gold or steel wire is recommended for use as oilers.

Naturally, the thicker the wire and its duck-billed end, the more oil it will hold. Therefore, the oil applicator should be judiciously selected so that too much oil will not be applied to the jewel holes. Too much oil will cause as much trouble as an insufficient amount.

One of the most important and skillful operations in watch repairing is the proper oiling of the movement. Therefore, it should be done with the utmost care and thought.

Begin assembly by lubricating the mainspring before it is wound in the barrel. A simple method of doing this is to soak a piece of tissue paper in clock oil. Fold this tissue over the spring, holding the fold with a pair of tweezers. Draw the spring through the soaked paper, pulling it from the end of the spring towards the inside coil until the entire spring is oiled.

Insert the mainspring into the barrel with a winder. (See chapter "Mainspring Repairs.") Place the barrel arbor in the barrel and replace the barrel cap. Oil the barrel arbor at the shoulders where they emerge from both sides of the barrel.

Cap Jewel Screws

Replace all cap jewel screws. Select the proper cap jewel screw for each cap jewel by observing it for polish and length. Usually, jewel screws whose heads are polished are inserted from the top of the bridges. Jewel screws whose thread ends only are polished are inserted from the bottom of the bridges. Those that are unpolished are inserted from the dial side of the main plate. These are called lower cap jewel screws. Jewel screws that are very short are made so necessarily in order to prevent them from emerging through the movement into a low wheel or pinion. Typical of such a screw is the lower escape cap jewel screw.

Oil all capped jewels by placing the oil into the hole jewel cup. To make certain that the oil will flow through the hole onto the cap jewel, take a fine-pointed gold wire with a diameter much smaller than the hole in the jewel. Insert this wire through the hole jewel. Capillary action will draw the oil onto the cap jewel and form a small ring of oil (Figure 17).

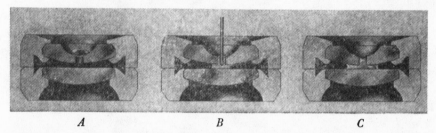

A   B   C

FIG. 17. *A* shows glob of oil on hole jewel. *B* shows thin gold wire forcing oil through to cap jewel (endstone). *C* shows how capillary action keeps oil around hole and cap jewels.

Some watchmakers prefer to oil the cap jewels just before they are replaced. This method is good because it does away with the necessity of forcing the oil through the hole jewel in order to make it contact the cap jewel. Its slight drawback is chiefly that it necessitates the additional oiling of the hole jewel. This additional oiling may result in adding too much oil and thus flooding the oil cups of the jewels.

If the jewels have been properly cleaned and if they appear clear, the oil may be easily observed. If the jewels are properly set, parallel to one another, capillary attraction will draw the oil onto the cap jewel and form a small bead in the shape of an oil ring (Figure 18). The oil ring should be larger than the cup in the hole jewel. This can be observed with an eye loupe, looking into the cap jewel.

The formation of an oil ring is an assurance that the bead of oil is contained around the hole in the jewel and that it has not spread and spilled away. All cap jewels should be inspected through the loupe after oiling to observe that the correct amount of oil has been applied, judging this, of course, by the size of the oil ring, which should be about ⅔ or less the diameter of the cap jewel itself.

The train wheels are then assembled, placing them into the movement in exactly the reverse order in which they were dismantled; the wheel that is lowest (and was removed last) in the train is replaced first. The wheel next highest is then replaced. In our model, the escape wheel is replaced first, fourth wheel next, then the third wheel followed by the central wheel. In this way, no wheel will have to be edged under another. The train bridge is then replaced, and, after making sure that all the wheel pivots are securely in their jewel holes, the center wheel is nudged with the tweezer to determine whether each wheel is placed correctly and is not binding.

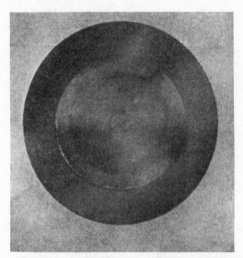

Fig. 18. The bead of oil is shown centered around the hole and occupies three-fourths of the center surface of the cap jewel.

Only a slight nudge of the center wheel should be sufficient to send all the train wheels into a smooth and easy rolling motion.

The barrel and bridge, ratchet, crown wheel, and click are then assembled, making sure that the crown wheel screw, which is shorter than the ratchet wheel screw, is not interchanged. Before assembling the ratchet wheel, the barrel arbor bearing in the barrel bridge is oiled lightly. The crown wheel washer is also slightly oiled. As the screwdriver fastens the ratchet wheel, the mainspring might wind up a slight amount, but it should be sufficient to send the train wheels into motion. During this time, observe the motion of the escape wheel. If it comes slowly to a stop or even recoils, it is a good sign that the train is clean and free. If the escape wheel stops abruptly, there may be some obstruction in the train or the mainspring is "set." A mainspring that came out of the barrel in a wound-up condi-tion similar to a hairspring should be discarded and replaced by a new one. Such a mainspring has lost its resiliency and has become "set."

Oil the pallet jewel impulse surfaces (Figure 19) and assemble. As-certain that the pallet is free and drops from one banking pin to another by its own weight without power from the train. Endshake and sideshake in the escapement must be provided for and checked.

Assemble the winding and setting parts in the movement. Oil the parts with a good clock oil or white vaseline. Check the winding and setting mechanism and make sure that the winding pinion and clutch wheel are properly engaged only *after* these parts are lubricated.

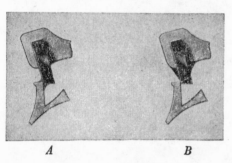

A   B

FIG. 19. A: Properly oiled. B: Too much oil.

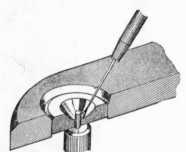

FIG. 20. How the oiler applies oil to a pivot.

Oil the center wheel post, lower center pivot, and lower third wheel pivot. To oil these parts, the oiler is applied to the pivot at the spot it emerges from the jewel. In this way, the capillary action will keep the oil around the pivot and jewel hole. This is illustrated in Figure 20.

Press the cannon pinion into position on the oiled center post. The can-non pinion must be replaced before the minute wheel, otherwise the cannon pinion teeth may nick the soft, brass, minute wheel teeth. If the cannon pinion appears to be too loose, it may be tightened using the methods de-scribed in the chapter, "Adjustments to the Cannon Pinion."

Assemble the dial train and setting parts. Of these the stem is lubricated by a light smear of clock oil or vaseline on the square section. The clutch teeth of the winding pinion and clutch wheel are similarly lubricated. The setting lever is oiled lightly at the point where it makes contact with the clutch lever. The setting bridge spring or "locking arm spring" is lightly oiled at the spot where it contacts the thin pin at the rear of the setting lever.

The oiling of the upper pallet hole jewel is a controversial topic in modern horology. However, a most recent survey indicates that this pivot should be slightly oiled. This is done by first dipping the smallest oiler in the oil and touching it to the fingernail or to the surface of a piece of tissue paper. The remaining oil is then applied to the pivot and jewel. In this way, there is no danger of superfluous oil overflowing onto the pallet.

Assemble the balance wheel in the watch and check it for endshake and balance concentricity. The hairspring must be properly centered and level; the regulator pins must be straight and parallel. There must barely be room between these pins for the hairspring to vibrate.

Check the balance assembly to see that it is in beat. A watch is in beat when the roller jewel holds the pallet fork midway between the banking pins without power in the train. If the fork is obstructed from view, wind the watch just enough so that the escape wheel can move. If the escape

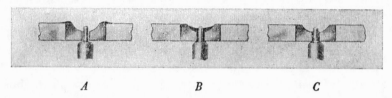

| A | B | C |

FIG. 21. Oiling train jewels. *A*: Improperly oiled. *B*: Too much oil. *C*: Properly oiled.

tooth falls upon the lifting surface of the pallet jewel, the watch is in beat. If the escape wheel tooth locks upon the pallet jewel, the watch is out of beat. For instructions on how to place a watch in beat, refer to the chapter "How to Place a Watch in Beat."

Oil the rest of the train wheel pivots, using the method shown in Figure 20. When oiling the jewels, make certain that the oil is confined to the cup in the jewel. Otherwise it will spread onto the plates or bridges, drawing the rest of the oil with it. (See Figure 21.)

The watch should then be wound. If the escapement is in beat and all operations up to the present have been satisfactorily performed, the balance should start after the first few windings.

While the watch is running, place a light drop of oil on every third or fifth tooth of the escape wheel. Make sure that the oil is applied to the *tops* of the teeth. Because some watches are constructed so that the escape teeth are not accessible to the oiler, this must be done before the pallet is assembled.

All screws must be checked for tightness, using a screwdriver of the correct width and sharpness. Make certain that long screws are not placed into holes where they might emerge from the other side and act as an obstruction.

The dial should be cleaned cautiously. Do not attempt to clean a dial with benzine if it has painted figures. Such a dial may be cleaned by dipping it for a few seconds in the cleaning solution, rinsing in water, and then drying it with a *blotting* action, using the Sylvet cloth. If the dial is not tarnished, it should then be brushed free of dust, using a soft brush. Porcelain dials may be cleaned by washing with soap and water, care being taken not to crack these brittle dials.

Assemble the dial train and hands to the watch. Make sure that the hands have the proper clearance from the dial and each other as well as following the curvature of the dial. The hands must be co-ordinated so that when the minute hand points to the figure "12" the hour hand is directly at the hour indicated and not between two figures. Make certain that the hands do not touch the crystal.

Clean the case of grit, dirt, and lint. Make sure that it is perfectly dry. Check the crystal for security in the bezel. On wrist watches with two piece cases, fit the movement into the case back. Place the bezel on the watch at the balance end of the movement and gently snap the other end shut.

Set the watch and make sure that the hands clear the crystal in all positions round the dial. Wind up the watch fully and listen to its ticks in all positions to make sure that the balance does not hit or scrape any part in the watch or that the case does not bind the balance or any other part of the movement and cause it to lose motion or stop altogether.

Summary

1. A watch movement should be cleaned and oiled like any other machine to prevent its wear and increase its efficiency.

2. The watch case should be opened with great care lest it be scratched or marred.

3. The movement is removed so that the balance does not become caught or broken on the case edge.

4. A piece of thin celluloid is used to protect the dial while the hands are pulled off with the hand remover.

5. The balance must be removed so that it does not get caught on other parts of the movement.

6. Release the power by pushing back the click with a piece of pegwood while the left hand controls the unwinding by grasping the crown.

7. The cannon pinion is removed by twisting it off the center post while it is held in the jaws of a pin-vise.

8. The pallet bridge, winding wheels, mainspring, and train wheels are dismantled in that order.

9. All cap jewels must be removed.

10. All parts are placed in a wire mesh cup, which is then placed into the cleaning solution.

11. The balance is cleaned first and dried, like other parts, by the blotting action of a clean Sylvet or linen cloth. Other methods of drying are evaporation by mild heat or blowing a jet of air from a watch blower.

12. The watch is assembled in the reverse order of dismantling; generally the first part removed is the last part assembled.

13. All holes and jewels must be cleaned with a pointed piece of pegwood, and cap jewels must be wiped clean with a chamois buff-stick or pegwood.

14. Oil should be applied to all jewels with endstones (cap jewels) so that there is a distinct oil ring around the hole in the jewel as observed through the cap jewel.

15. The winding must not be tested before it is oiled.

Questions

1. How often should a watch be cleaned and oiled? Why?
2. How is the movement removed from the watch case?
3. How are the hands removed and how is the dial protected during this operation?
4. What precautions are necessary when removing the balance wheel?
5. When is the power in the mainspring removed and how?
6. How is the pallet bridge removed to avoid bending, scratching, or loss of the pallet?
7. How is the cannon pinion removed?
8. How is the barrel arbor taken out of the barrel? How is the mainspring removed from the barrel? How is the barrel cover removed?
9. What is the procedure in removing the train wheels?
10. How are the watch parts cleaned? How are plates that are tarnished cleaned?
11. What are some safety precautions to be followed in the use of rinsing solutions?
12. What are some methods of drying watch parts?
13. How are the following oiled: cap jewels, train wheel pivots, pallet jewels, and the mainspring?
14. What revolving or moving parts in a watch are *not* oiled?
15. How is the movement replaced in the case to avoid damage to the balance?

III

HOW TO CASE A WATCH

Fitting a watch movement into a watch case is called casing. To do this job efficiently, the movement must receive the greatest protection from dust and moisture. The completed unit must present the best appearance and permit easy winding and setting of the hands.

All these considerations call for skill, knowledge, and a high regard for the minutest details.

The skills in watch casing fall into five main divisions:

1. Fitting dials.
2. Fitting and adjusting the hands.
3. Fitting stems and crowns.
4. Case reaming or scraping.
5. Reaming or filing the stem opening in the bezel and case bottom to suit the stem and crown.

All these operations may not always be encountered in one job but may occur wholly or in part in all casing jobs.

PART I: DIALING

When a new watch is made up or an old one remodeled, a new case is selected to suit the size of the movement. The bezel opening selected generally conforms to the popular style at the time. A new dial must be purchased to suit the opening in the bezel. Sometimes it is possible to refinish (repaint) the dial to match the new case. Importers of new movements and domestic manufacturers co-relate cases, designs of new movements, and

38

dials well in advance so that the dialing operation is unnecessary. However, sudden changes in styling and other unforeseen factors often make dialing a necessity.

In fitting a new dial to the movement, the following must be checked:

1. Bezel opening.
2. Dial leg position.
3. Position of hour and minute hand hole.
4. "Lining up" of the second hand hole.
5. Alignment of all these factors so that the figure "3" is exactly at the stem position.

Sometimes all these may be in alignment but the dial may not align with the bezel opening when the case is shut as shown in Figure 1. This error is due to faulty printing on the dial. Do not use such a dial. Attempting to align this dial would bring into practice methods not consistent with good watch making.

Dial legs serve two purposes. First, they steady the dial on the movement; second, they serve to keep the dial tight against the movement plate.

Figure 2 shows a dial hip screw. The sharp-edged hip has a section filed off to permit the passage of the dial leg. When the hip screw is turned halfway around, the sharp edge bites into the side of the dial leg and grips it securely. To release the dial, the screw is turned to the open position (Figures 2a and 2b). Another type of dial leg fastening is shown in Figure 3. This shows a "side

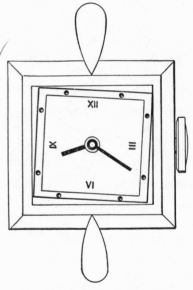

Fig. 1. A dial out of alignment.

dial screw" set into the edge of the movement plate. This is screwed in until the pointed tip of the screw engages the dial leg securely.

Occasionally, dial legs are too thick to fit into the holes in the bottom plate. It is then necessary to file the legs thinner. This can be done with a small, square needle file. The dial is held at the edge with one hand and the dial leg is reduced by a series of short, circular strokes with the square file. This is illustrated in Figure 4. Thinning a dial leg may also be done by using a small pinvise. Insert the dial leg loosely and rotate the pinvise.

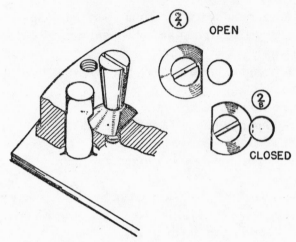

FIG. 2. How a "hip" dial screw secures the dial leg.

The sharp corners of the hardened jaws will shave off sufficient metal to thin the dial leg to size.

Dial legs should not be longer than needed to be engaged by the dial screw. Dial legs that are too long may extend into a wheel, stopping the watch, or butt against the bottom of one of the bridges. This prevents the dial from lying flush against the bottom plate.

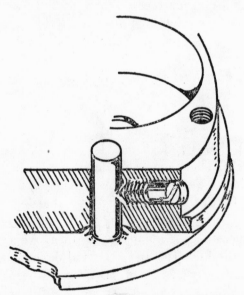

FIG. 3. How a side dial screw secures the dial leg to the movement.

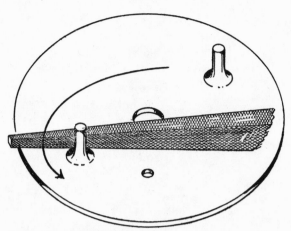

Fig. 4. How a dial leg may be made thinner with the circular motion of a square needle file.

Occasionally, the dial will refuse to remain flush with the movement
even when the legs are corrected to size. This will prevent the proper closing of the case as well as cause the undersides of the hands to scrape the
dial. The dial may not set flush with the movement because the coned
section at the base of the leg is too thick (Figure 5). This coned section
should be reduced by running the square needle file around the base of
the dial as shown in Figure 4.

When the dial is placed on the movement, the hour wheel tube must have
sufficient clearance around it in the dial hole. The same is true of the
second-hand post. This clearance is necessary so that, when the hour hand
is placed on the tube, the sides of the hand socket will not touch and bind
against any side of the dial hole. If this goes undetected, the watch may
stop.

In Figure 6 the arrows point to the parts of the hour-hand socket and
second-hand socket touching the sides of the dial holes. This may be the
fault of the dial legs being bent. The dial legs must be straightened before

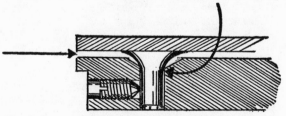

Fig. 5. Because the dial leg is tapered excessively, it keeps
the dial from lying flush against the plate.

placing the dial on the movement. Bending the dial legs purposely to attempt correction of the faults shown in Figure 6 may correct this condi-

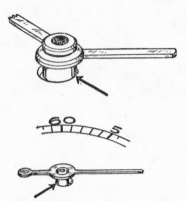

tion but it also may introduce an alternate error of the dial being out of line as illustrated in Figure 1.

If the holes are too small to provide freedom or clearance for the hand sockets, they may be enlarged. This is done with a broach or a round needle file. The file is inserted through the back of the dial, a precaution taken so that, should the file slip, it will not mar or scratch the face of the dial.

If the hole touches on one side only, the fine needle file is used. The file is made to bear against the side of the dial touching, say, in the direction of the numeral 3. It is filed just enough to provide the clearance needed. An excessively large hole in the

FIG. 6. When the dial is not centered or its legs are bent, the holes may bind the hands (shown by the arrows) and stop the watch.

dial is unsightly and provides an entrance for dirt and lint. The hole should be made to appear round by streamlining the part filed with the rest of the hole. Remove all burrs from both sides of the dial with a fine countersinking tool.

Dial Washers

Dial washers are wafer-thin copper discs holed in the center to fit over the hour wheel. They are rolled to make them bend up at the sides and become springy. Washers are used when the endshake of the hour wheel

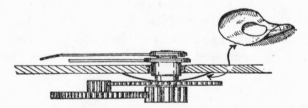

FIG. 7. How a dial washer keeps the hour wheel enmeshed
with the minute wheel pinion.

under the dial is so great that it may become disengaged from the minute-wheel pinion. The dial washer, placed over the hour wheel tube and under the dial, keeps the hour wheel away from the dial and engaged with the rest of the dial train.

Dial washers are not necessary when the endshake between hour wheel and dial is at a minimum or the amount of endshake is less than the distance from the bottom of the hour wheel to the top of the minute-wheel pinion when they are engaged. This minimum endshake may be determined by grasping the tip of the hour tube that extends through the dial with the tweezer, lifting it upwards, and attempting to turn it. If the wheel has become disengaged, it will turn with the tweezer. A similar test can be applied after hands are secured by lifting the hour hand. If it can be turned to another position without effort, then the wheel has become disengaged, indicating the need of a dial washer. A dial washer and its use are shown in Figure 7. Dial washers should not be used unless all other means for correcting the situation are unavailing.

PART II: FITTING THE HANDS

Hour, minute, and second hands must fit in such a manner as to give a clear-cut appearance and facilitate quick and accurate time reading. The hour hand must fit the hour wheel tube so that it goes down flush with the top of the tube or, in special cases where the hour wheel has a shoulder, down to the shoulder. Minute hands, likewise, should fit tightly and flush with the polished tip of the cannon pinion.

FIG. 8. Showing the proper proportions for clearance of the hands with the dial.

The second hand must fit snugly and clear the dial with a minimum space and yet not interfere with the hour hand or the tip of the minute hand.

Watches with flat dials should have their hands fitted so that they approximate the tolerances indicated in Figure 8. Hands in curved dial watches should be fitted with a minimum tolerance in order to present a thin, streamlined effect.

When the hands are fitted to the movement, the hour hand must have a slight but noticeable endshake (up and down) and a slight sideshake. Thus the tolerances mentioned above will be provided which will insure that the hands will not touch one another at their sockets and that the hour wheel tube will not pinch or bind on the cannon pinion.

Tightening Hands

When a second hand is loose on its pivot, it may be tightened by placing the tube of the hand in a pinvise or lathe chuck and closing the tool slightly. This tightening closes the hole in the second-hand tube sufficiently to become friction tight on the fourth wheel pivot. The operation is shown in Figure 9. Do not force a tight second hand on a pivot as this may cause the pivot or the jewel of the second post to break.

FIG. 9. How hands of a *bombé* dial may be curved for streamlining and clearance.

FIG. 10. A method of closing the hole in a seconds-hand tube.

Minute hands that are slightly loose may be tightened by using the methods shown in Figures 11 and 12. Figure 1 shows the socket of the minute hand resting on a steel block with a round bottom punch over it. A few taps of the hammer should close the hole enough to secure the hand to the cannon pinion. Figure 12 shows the bell punch constricting the hole of the socket to make it smaller.

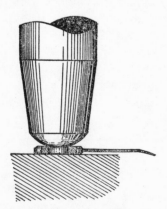

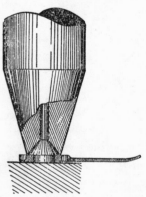

FIG. 11. Reducing the hole size in a minute hand.

FIG. 12. Contracting the hole in a thin socketed minute hand.

Hour-hand sockets are made smaller with the
bell or convex punch used as shown in Figure 13.
Here the punch is brought over the hour-hand
socket. By tapping the punch lightly, the conical
section of the punch contracts the short tube of the
hand enough to give it a firm hold on the hour
wheel tube.

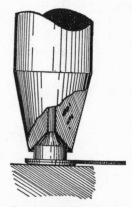

Broaching Hands

When the holes in the hands are too small, they
may be opened by broaching. Broaches are small,
tapered needles with five or six sharp edges. The
hands to be broached must be held in a manner that
will grip the sockets of the hands and thus prevent
breakage.

Fig. 13. Constricting the
hole in an hour hand.

A simple method of holding the hands is shown in Figure 14. This tool
is called a hand broaching vise. The hand is placed over the hole that is
slightly larger than the hole de-
sired. The top disc closes over the
hand and is screwed tightly shut.
The broach is placed through the
top and, when it is twisted, the
stem of the hand is free, the hand
being gripped at the socket, thus
relieving any strain. Another
method is shown in Figure 15. A
tweezer with several notches of
different sizes is used. The hand
to be broached is held in the open
corners of one of these notches.
When the broaching operation
takes place, the strain is upon the
socket, leaving the extended hand
free. Both these tools are obtain-
able at material jobbers.

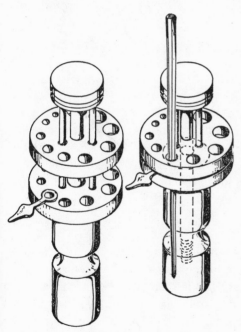

When the hands are placed on
the watch, the hour and minute
hands should be co-ordinated.

Fig. 14. An effective hand broaching vise.

(See Figure 16.) It is a good practice to place the hour hand on the numeral "3" and the minute hand on the numeral "12." This permits rapid

checking and provides plenty of room in fitting the second hand. Hands not placed precisely result in confused time reading as shown in Figure 16 where the hour hand is between the numerals "3" and "4" while the minute hand denotes the hour.

When the hands are in place on a flat dial, they should clear one another as shown in Figure 8. The tip of the hour hand should clear the second hand. The hour hand must be flat and parallel with the dial. The minute hand, however, should be bent downward near the tip so that the bottom of this hand is on the same level or plane as the hour hand. To bend this hand further may cause interference with the second hand.

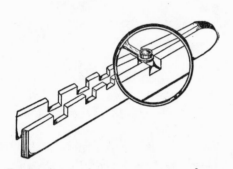

Fig. 15. A pair of strong tweezers may be converted to hold hands while they are broached.

Fig. 16. Hands set out of coordination.

The minute hand is generally bent at the tip to prevent it from touching the crystal. If it is suspected that the minute hand does touch the crystal and it is not possible to bend the hand lower, it may be tested. This is done by placing a small bit of oil on the part of the hand suspected. Should the hand touch the crystal, an oil smear will appear on the crystal when the minute hand is moved.

Length of Hands

The proper length of the hands is an important factor in providing good appearance and ease in time reading. Hands that are too long usually defeat their own purpose by obscuring the numerals. They also destroy the harmony of a well-proportioned dial. Hands that are too short make for guesswork in time reading. Again, hands that are too long may catch against the edge of the case and stop the watch (see Figure 17).

A minute hand should be just long enough to reach the outer edge of the numerals or the minute track. Hour hands should be about three-fourths the length of the minute hand. Sweep second hands should be long enough to reach to the seconds track or seconds division on the dial. The proper length of hands is shown in a cut view of a sweep second dial and hands in Figure 18.

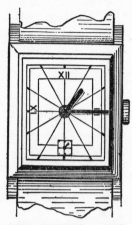

FIG. 17. Hands must be of proper length, or they may catch on the edge of the case.

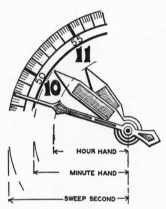

FIG. 18. The correct proportionate length of hands.

In oval or rectangular dials or those other than round, the minute hand should not be longer than the outside edge of the figure nearest the center of the watch.

PART III: FITTING THE MOVEMENT INTO THE CASE

A case of fine quality and finish needs little adjustment to make it accommodate the movement accurately. Occasionally, cases are just a bit too tight for the movement. A well-fitting case receives the movement without too much pressure, yet holds the movement securely. A case that permits the movement to twist or shake up and down is not desirable.

When a case is too tight for the movement, some casers take a metal block of the same shape and size as the movement and force it first into the case. Then they tap the top edges of the case with a horn head hammer. This either stretches the case bottom or corrects any tiny kinks.

If the tapping fails, a case scraper is used. In round cases, a scraper is used to scrape off a thin strip of metal evenly around the inside of the case, making the diameter just a bit larger to permit a snug fit.

In odd-shaped case bottoms, scraping should be done only after accurately determining the exact parts to be scraped. Care should be exercised

not to distort the case. When scraping is done haphazardly, the result may
be as indicated in Figure 1. Scraping a case is shown in Figure 19. Here
a rectangular case is being scraped at the corner. The scraper in this case
is a thick hand-lathe graver set into a handle and sharpened to a keen edge.
This tool gives good service and provides a clean cut without danger of
scratching the bottom of the case. The case is held rigidly and securely.
Control of the graver or scraper must be maintained while pressure is
applied.

FIG. 19. How the case is held during case scraping.

All chips and burrs should be removed. The scraping should be done
so that the result is a neat, clean and even-surfaced edge. It is seldom
necessary to scrape cases excessively. If the movement is very loose in the
case, little can be done unless a case ring or washer is used to fill the space
between the movement and the case.

When the case is slightly loose, the slack may be taken up by gently
bending the metal near the stem opening inward. This applies to two-piece
cases only. Care should be taken not to mar the appearance of the case.

The movement should be handled with care while placing it into the
case. Otherwise the balance may be injured by contact with the case edge.
Therefore, the balance side of the movement should be inserted first. The
rest of the movement may be snapped into the case if it does not require
too much pressure. Failure to observe this safety precaution often results
in the breaking of the balance staff. The stud screw and the barrel are two
other parts of the watch that may extend beyond the edge of the movement
and so precaution must be taken to prevent these from catching on the edge
of the case as well.

Fitting the Stem

After the case is adjusted to receive the movement and the dial and hands
are fitted, the stem is shortened to a length best suited to lend beauty to the

watch. It should also prevent dust from entering the opening in the case and permit ease in winding and setting.

New movements are supplied with stems of sufficient length to suit practically all types of cases. These stems are filed to the effective length. It is a good practice, before removing the stem from the movement, to pull the stem to the hand-setting position. Doing so forces the clutch wheel snug against the intermediate wheel and prevents it from slipping off the clutch lever when the stem is out. When the clutch slips out of the clutch lever, it is often necessary to remove the dial and hands to reset these pieces.

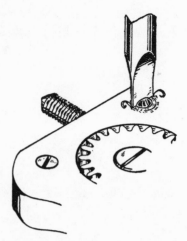

Fig. 20. Using screw drivers of improper size will score the movement.

To remove the stem after it is pulled into the setting position, loosen the setting lever screw just enough to permit extraction. Unnecessary turn-

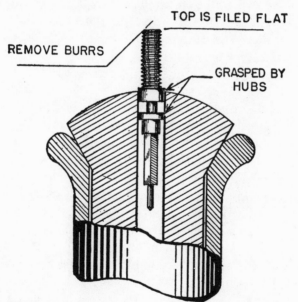

TOP IS FILED FLAT

REMOVE BURRS

GRASPED BY HUBS

Fig. 21. The proper method of holding the stem while the thread length is prepared.

ing out of the setting lever screw will cause the setting lever to fall off, again necessitating removal of the dial and hands for this replacement.

Screwdrivers must be sharp and fit the screw slot to its depth. The width of the screwdriver blade must not be wider than the screwhead. A screwdriver blade that is wider than the screwhead will score the metal surrounding the screw and mar the new appearance of the watch. This is illustrated in Figure 20.

When the stem has been removed, it is placed in a pinvise and cut to approximate length with a cutting plier. When the stem is to be filed to length in a lathe chuck, it is cut before "chucking" the stem.

The stem should be held in the chuck of the lathe or pinvise so that it is held by both hubs. This diminishes the chance of the stem breaking at its weakest point—the setting lever slot. The proper method of grasping the stem is shown in Figure 21.

Length of Stem

When a stem is fitted with a crown, the bottom of the crown should barely clear the edge of the case, leaving enough room for the fingernail to lift the crown for the setting of the hands. (See Figure 22.) In this way,

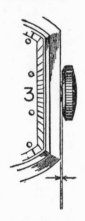

the neck of the crown fits snugly into the case hole opening; keeping out excess dust or moisture. A crown that protrudes excessively exposes itself to breakage as well. Conversely, a crown that is so short that it rubs against the case makes winding and setting difficult. The rubbing of the crown against the case may prevent complete engagement of the winding wheels and result in a poor winding condition.

Crowns are usually supplied with the case. However the crown should harmonize with the design and predominant color of the case. It should be wide enough to allow easy winding. A crown that is too small will prevent the watch from being fully wound with the subsequent complaint that the watch will not run 24 hours.

FIG. 22. Proper clearance of crown with case.

The stem should be filed to receive the crown, using a fine file. Files that are too coarse may catch and break the stem. The tops of the threads should be flat, with the burrs removed as shown in Figure 21. This will permit the crown to be fitted and fastened securely, flat, and without the possibility of becoming loose. The crown should be fitted to the stem so that it is flat and will not wobble when turned. (See Figures 23A and B.)

If the crown resists efforts to be secured perfectly flat and is just slightly off angle, it may be trued by tapping the high side of the crown with a soft horn hammer. When this is done, the crown must be fastened again.

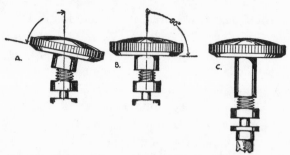

FIG. 23. *A*: Crown improperly secured. *B*: Crown fastened correctly. *C*: Long neck crown.

To test the truth of the crown before inserting it into the movement, roll the pinvise on the bench. If the crown is true, it will not wobble while the pinvise rolls. When the lathe is used, the stem is rotated slowly for this test.

Before replacing the stem and crown into the movement, it should be lubricated by dipping the tip of the stem into oil. The oiling makes the winding and setting lighter.

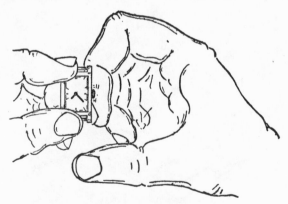

FIG. 24. Testing ease of winding.

To replace the stem and crown, it is eased gently into the movement with a slight twisting motion so that the square of the stem is "keyed" into the square of the clutch wheel. The setting lever screw is then tightened slightly. The stem is pulled out a little to permit the setting lever pin to fall into the setting lever slot of the stem. The screw is then fully tightened and the stem put into the winding position and tested.

The movement is then inserted into the case and the length of the stem and crown observed. If it is too long, the stem is removed and readjusted, repeating the operations listed above. If the stem is too short, a new one will be needed. Sometimes, however, it is possible to use a crown with a long post to take up the needed length. This must answer all the requirements of a well-fitting crown. Such a crown is shown in Figure 22c.

When the stem and crown are of the proper length and the movement placed into the case, the top is snapped shut. Then the ease of winding and setting is tested. A well-fitted stem and crown, set into the case, should have enough freedom in the case hole to permit winding and backwinding with ease. Backwinding is the opposite movement of the crown in the direction of winding. Backwinding is tested by placing the front side of the index finger against the side of the crown and winding it back and forth with only slight pressure against the crown as shown in Figure 24. If the crown skids against the finger, it may be an indication that the hole in the case is causing the post of the crown to rub against the side of this case hole. This hole must then be enlarged or filed to suit the crown post.

Filing the Hole in the Case

To determine where the case hole must be enlarged, pull out the crown into the setting position. This will permit visual observation with an eye

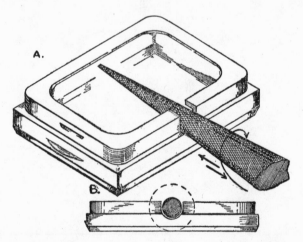

Fig. 25. Enlarging the hole for the stem and crown neck.

loupe. Observe where the rubbing takes place or at what point the case hole touches the crown post. This contact may be against only one of the sides, the top of the case, or the bottom of the case. The hole may also be entirely too small for the stem and crown, which would prevent proper closing of the case.

To enlarge the hole, a fine, round needle file is used. The stroke of the file should be short but deliberate, using a forward twisting motion bearing against the side to be filed. This is shown in Figure 25A. The shape of the hole and the clearance for the crown is indicated in Figure 25B. Metal and dust particles must be removed from all crevices and corners of the case. This is done with a blower and a fine brush. The case should not be scratched during any of the foregoing operations.

Waterproof Watch Cases

There are four types of waterproof watch cases. 1—Screwback; 2—Snap-back; 3—Push-out; 4—One-piece. The screwback shown in cross-section in Figure 26 has on its back, flanges, serrations or slots to accommodate a wrench. Use a sturdy, tight fitting wrench to open and close these. Figure 27 shows a snap-back type. Use a case knife as a wedge inserted against the lip or slight groove provided. Do not pry the case open. Figure 28 shows the push-out type and Figure 29 shows how this is done. The "push-out" is also recognized by the U-cut on the bezel to accommodate the crown-pipe. The one-piece case has none of the preceding three type features. This type has but a one-piece case in which the movement rests. The crystal keeps the movement in place. The movement is removed by constricting the plastic crystal with a special constricting pliers or wrench. The crown and stem are of the snap type (Figure 30). This unit and its plier are shown in Figure 31.

Movements which are removed through the front of the waterproof case employ the snap-stem and crown. These are removed by exerting more outward pressure than is required to set the hands. A method of removing such crown is shown in Figure 32.

In casing or replacing movements in waterproof cases, make certain that the movement is secured and the stem and crown are in perfect alignment. The gasket should be clean and unwrinkled, not frayed or aged. The use of silicon grease on top of the gasket and in the crown aids in waterproofing and helps in tightening the back without crimping the gasket. If a waterproof watch case shows condensation it is positive proof that there is moisture in the watch and that the case is not waterproof. A simple test for waterproofing is shown in Figure 33. Place the watch in a glass of hot water (120°–140°F). The heated air in the case, expanding will escape through the leaky spots and this will be observed by the bubbles.

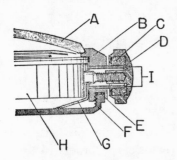

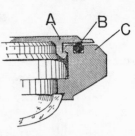

Fig. 26. A typical waterproof watch case. A—Crystal; B—Case frame and main body of case. C—Gasket of rubber, plastic or cork. D—Crown. E—Gasket for case back. F—Screwback G—Movement tension washer or ring. H—Movement. I—Case-pipe over which crown and gasket fit.

Fig. 27. Typical snapback case construction. A—Snap-back. C—Ring gasket of plastic or rubber. C—Main case body.

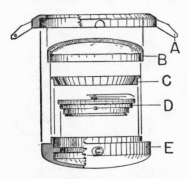

Fig. 28. The push-out case. A—Top bezel. B—Crystal. C—Reflector and crystal reinforcement. D—Movement. E—Case bottom. This uses the snap-stem and crown.

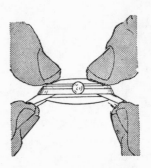

Fig. 29. To dismantle the three-piece case, hold watch as shown and press with thumbs.

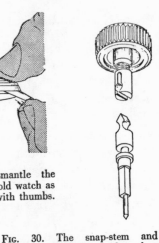

Fig. 30. The snap-stem and crown, used only in watches when movements are removed from the front, is strong enough to be pulled out into the hand setting position without disconnecting. To remove from case and movement, see Figure 32.

Fig. 32. A safe method of removing a snap crown: plastic foil and dull pliers. Do not pinch case-pipe.

Fig. 31. An exploded view of one-piece case. Pliers at top constricts crystal to enable movement to be removed after snap-crown is disengaged.

Fig. 33. A simple test for waterproofness—equal to four foot depth of water—consists of dipping watch in water between 120°F and 140°F. Remove immediately when and if bubbles appear (at leaky spots).

Fig. 34. A pressure tank to test watches in pressures up to 57 lbs. p.s.i. or equal to a water depth of 131 feet.

Self-Winding Watches

In casing self-winding watches, make certain that the oscillating weight does not bind or scrape the inside of the watch case. This difficulty may arise if there is a loose movement, a missing gasket or one too thin. Test by placing a light drop of oil on the weight, closing the case and twisting to activate the weight. If the weight touches, the oil will be transferred to the inside back.

QUESTIONS

1. What are the requirements of a well-fitting watch case?
2. What are the requirements of a dial to be fitted to a movement and watch case?
3. What are the two main types of dial screws and how do they work?
4. How is a dial leg made thinner?
5. What is the purpose of a dial washer? When is it used?
6. How are the hands tested for clearance and freedom from binding?
7. How is a loose minute hand tightened? A loose hour hand tightened? A loose second hand tightened?
8. How are the holes in watch hands made larger? What precautions must be observed to prevent the hand from breaking during this operation?
9. What is meant by "blocking" a watch case?
10. What is the proper length of a stem and crown?
11. How is the crown tested for truth?
12. What precautions must be observed everytime the stem is removed from the movement?
13. What tools are used to make a case larger in order to receive the movement?
14. What is a test to determine whether the stem and crown are free from binding?
15. How is the case cleared of filings before the movement is inserted?

IV

ADJUSTING CANNON PINIONS

P ractically all adjustments needed by the cannon pinion result from
its becoming loose on the center post.

The obvious symptom of a loose cannon pinion is that it does not grip
the center post with sufficient tension to be carried around with this wheel-
post. Without this tension it cannot turn the minute and hour wheels.
Failure is almost always due to lack of proper lubrication, especially
with the excessive hand-setting on some calendar watches. Another symptom is that the watch loses considerable time or the hands remain motionless although the balance motion is good. Winding is normal, but while
in the hand-setting position the hands require no effort to be moved.

The adjustment needed is to restore this clutch-traction which originally was supplied by a nick or dent in its neck. This enables it to be
snapped onto the center post and grips this so that it turns along with
this post and in turn drives the dial train and hands. Tightening this too
much will make hand-setting too stiff with danger of shearing off the dial
train teeth.

Methods of Restoring Clutch-Traction

There are two main methods of restoring this clutch traction. One is
to renew the crimp or dent; the other, developed by this author, press-
spins a light groove around the neck, thus allowing it to grip the arbor-
post evenly all around. This latter system is simpler, quicker and good
for all clutch traction pinions and is particularly suited to those very
short cannon pinions used in off-center drives for the hands. Both
methods will be illustrated and explained with the press-spin system

employed to demonstrate adjustments to the very short cannon pinion repair.

Removing the Cannon Pinion

If the cannon pinion resists removal with ordinary methods, place the cannon pinion in a pinvise as shown in Figure 1. Twisting this upwards with a pulling motion should remove this easily. This method prevents crushing the cannon pinion which may occur should a cutting plier, for example, be used to pull this out.

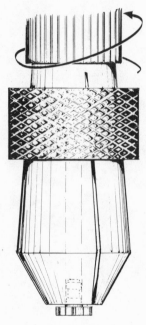

Fig. 1.

Renewing the Dent

Figure 2 shows the traditional method of restoring the tension to cannon pinions. A "V" stump is placed in the staking set, centered and locked in that position. A tapered broach or soft brass taper pin is placed through the pinion. If a broach is used, place the pinion so that the flat top of the broach is uppermost in the pinion as shown in Figure 2. The taper pin or broach prevents crimping the pinion-neck too deeply. Next the pinion is rested in the V of this stump and either a prick or

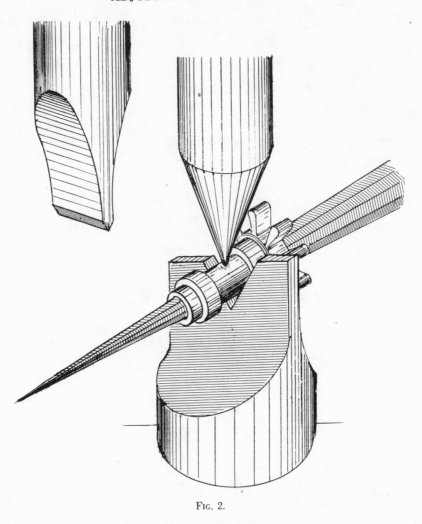

FIG. 2.

chisel punch is made to rest on the exact spot desired, or in the old dent. If the old dent appears worn or weak, place the punch at the same circumferential level. A slight tap with the hammer is all that is needed to make or renew the dent.

A properly placed dent is shown in Figure 3A. If this dent is placed too low, for example, it will cause the cannon pinion to ride up the center wheel post as shown in Figure 3B; or it will be too loose, as the dent will be opposite the groove in the post where it is deepest. This may result in failure of this pinion to enmesh with the minute wheel and permit the hour wheel and its hand to move out of coordination with the minute hand

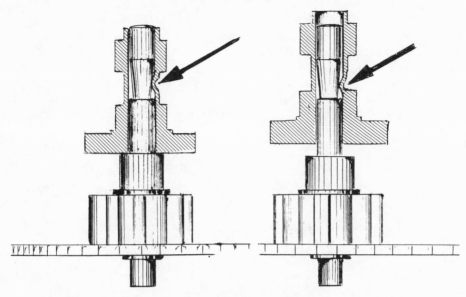

FIG. 3A (left) and FIG. 3B (right).

as shown in Figure 16, page 46. Or, it will be too loose despite the new dent.

Closed Top Pinions

When a closed top pinion requires adjustment, use a broach or brass taper pin whose tip is broken off so that its taper allows it to fill most of the area inside the pinion. This will give it protection against crushing during the nicking or spinning operation. This is shown in Figure 4.

Off-Centered Driving Cannon Pinions

Figure 5 shows a new type of train wheel arrangement which allows the seconds-carrying fourth wheel to occupy the center. This results in a different cannon pinion arrangement as well as larger barrels and balances. A is the barrel, B the main wheel (formerly the "center wheel"), C is the third pinion which is driven by B but C also drives the coupling wheel E, upon which the cannon pinion is frictioned. D, the third wheel, drives the centered fourth wheel pinion or "sweep second wheel." Traction of the cannon pinion is supplied by the nick in the back extension of the cannon pinion, shown by the arrow in the upper right inset of Figure 5. Do not separate E from F except to adjust the friction. Sometimes the cannon pinion and the coupling wheel are placed *between* the

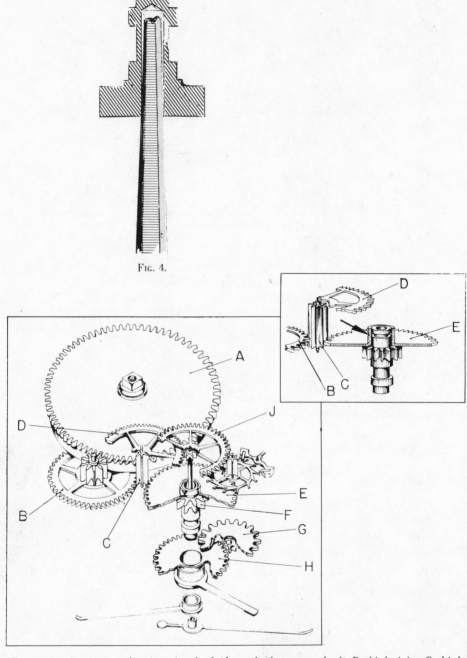

Fig. 4.

Fig. 5. Gnat's eye view of main train wheel (formerly the center wheel) B, third pinion C, third wheel D, and coupling wheel E. Friction is gained by a nick on the back tube at E at arrow point. G is the minute wheel and pinion. H is the hour wheel.

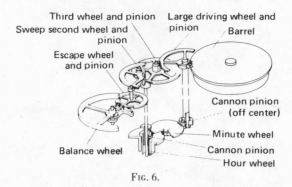

FIG. 6.

plates of the movement and can only be removed by dismantling the movement. Such watches have very short cannon pinions which do not carry the minute hand. Examine such watches carefully before attempting to pry off the cannon pinion.

Watches with Very Short Cannon Pinions

Figure 6 shows a form of train using two cannon pinions. Here the driving wheel's pivot doesn't extend beyond the dial plate. The cannon pinion snaps onto the grooved arbor before assembling as shown in Figure 7, making it appear to have two pinions in which the cannon

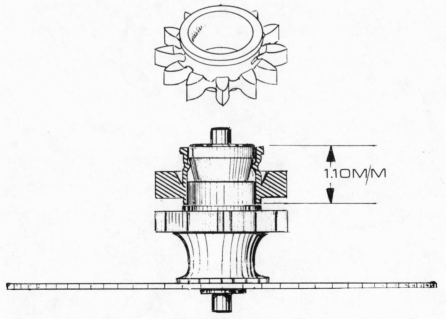

FIG. 7.

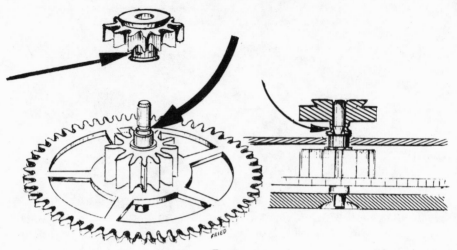

Fig. 8.

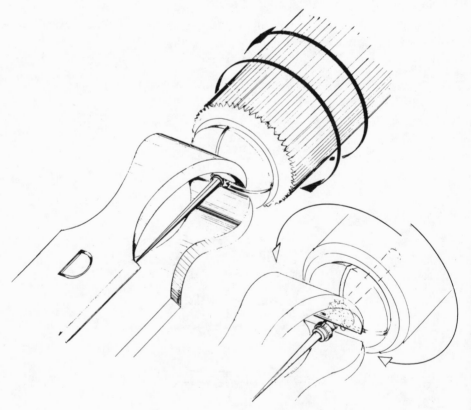

Fig. 9A (left) and Fig. 9B (right).

pinion itself enmeshes with the minute wheel through a side cutout in the dial plate. Others, for example, as shown in Figure 8, are attached onto a short post which emerges through the dial plate and is snapped onto this post in an up-side-down position. In many of these watches, the cannon pinion which holds the minute hand is centered and loose. This MUST NOT be tightened. Friction loss is due most often to neglectful lubrication to the frictional surfaces of the cannon pinion.

Tightening Short (Off-Center) Cannon Pinions

Remove the cannon pinion carefully, using D-tweezers and placing a finger over the pinion to prevent its loss. Place a soft brass taper pin through the pinion side and place the pinion in a pinvise so that its leaves are grasped by the jaws of the pinvise. Lubricate the outside surface of the cannon pinion's neck with stem grease. Next use a D-tweezer to grasp the extending neck of the pinion at a point where the original friction-dent was placed.

While pressing the D-tweezer jaws moderately, twist the pinvise back and forth about three times as shown in Figure 9A and B. This causes the D-tweezers's sharp jaws to spin a groove around the neck of the cannon pinion as shown by the arrow in Figure 10. This system is good for both long and short pinions. On longer pinions shown in Figure 9B it is better to use a soft brass taper pin insert. The spin-pressure depends on the size and thickness of the cannon pinion tube. Do not place the pinion in a lathe as the tendency is to over-spin with the possibility of shearing the pinion tube.

Replacing a Cannon Pinion

If a cannon pinion is snapped onto the center wheel carelessly, its steel leaves may squash the softer minute wheel teeth. This often escapes detection and may cause the watch to stop after about three hours. As soon as the hands are reset so that the broadened minute wheel tooth is out of the cannon pinion, the watch starts off again until the same period expires—and so does the watch. To insure against such damage, remove the minute wheel before replacing the cannon pinion. Another possible point of damage is the upper center jewel. If the cannon pinion is pressed on without backing up the lower center pivot, the center pinion may push out or crack the lower center jewel and also bend that bridge.

Figure 11 shows the proper method of replacing a cannon pinion (minute wheel removed). A punch or stump with a concave surface whose tapered diameter will allow the back pivot to rest without interference with the plate or jewel is chosen. Line up the stump in the staking set and lock it. Then a flat faced, hollow punch is chosen which closely fits over the cannon pinion. The movement is positioned as shown, held in one hand; the punch merely pressed over the cannon pinion with the other hand should seat the pinion. Of course, first oil the post of the arbor before pushing on the pinion.

Testing a Cannon Pinion's Tension

Before replacing the minute wheel, test the pinion to see whether the pinion is still too loose or too tight. This is done by again inserting the pinion while on the movement into the jaws of a pinvise and twisting it

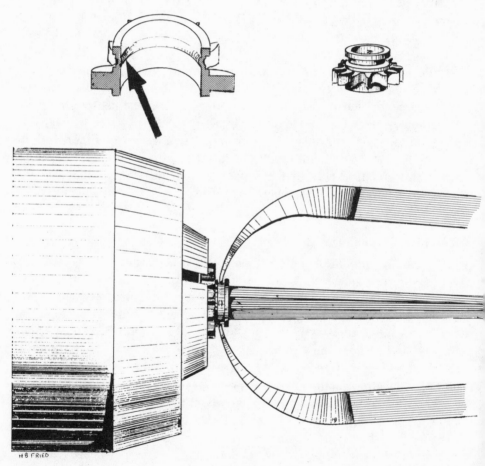

Fig. 10.

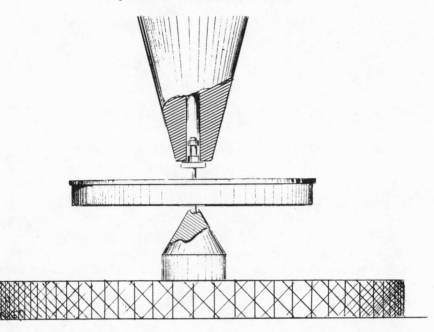

FIG. 11.

on the post. Some pinions may require great pressure to "snap" them
into place because of the crimp needing to overcome the full diameter of
the arbor. Then after it is snapped, it is loose. This is because the undercut or groove in the arbor is rather deep and the dent, spin-ring or crimp
doesn't yet contact the inner surface of the snap-groove in the arbor. In
such a case, the pinion must be removed and the crimp or groove made
deeper. This situation is similar to a finger ring that requires forcing
over the knuckle and then is loose below it.

If the pinion seems too tight its tension must be relieved, otherwise
the hand-setting operation may be too difficult for the wearer or the
minute wheel teeth may be broken when setting the hands against the
rigidity of the pinion. To remove some of this tension, place a tapered,
oiled steel pin through the cannon pinion. Twisting this deeper into the
pinion's hollow tube will force out some of the tensioning, yet not erode
the inner wall or the crimping. A correctly tensed pinion is when the
drag of the pinion is felt as the crown is turned when setting the hands. In
slightly turning the crown back and forth the sideshake of the minute
wheel teeth straddling the cannon pinion's leaves can be felt. It shouldn't
be so tight that hand-setting becomes uncomfortable.

Slotted Cannon Pinions

Older American and some Swiss watches sometimes used slotted cannon pinions. To restore their tension, merely force these splines back again using a round bottom peening punch as shown in Figure 13, page 263.

Coupling Clutch Wheels (with Cannon Pinions)

Figure 12 shows the method of restoring the clutch action to the coupling clutch with cannon pinion attached. This is also called a driving wheel arrangement. While most manufacturers advise a new unit, the clutch action can be restored. Separate the wheel from the pinion carefully so as not to bend the thin driving wheel. Place this on a flat stump D as shown and carefully center a bell-mouthed punch C whose diameter will fit midway over the three inner prongs. *Make certain that this punch rests concentrically* with the wheel and prongs. A slight tap is all that is needed to cause a minute concentric contraction that will restore the clutch action. Do not try to manipulate the prongs manually with a tweezer or other tool.

Replace the driving wheel on the back shoulder of the canon pinion, using a staking tool. Center the cannon pinion downward in a well-fitting hole of the staking die. Push the wheel onto the back shoulder of the pinion with a hollow flat punch just as you would position a snug balance onto a new staff. Lubricate the frictional areas of the clutch action on this and all types of cannon pinions.

Notice that in Figure 12, B represents a tube which is friction-fit into the plate and serves as the post upon which the cannon pinion and driving wheel combination rides. The tube also serves as a bearing for the sweep seconds pinion or centered fourth wheel. If the cannon pinion is tightened it will ruin the tube; the latter must then be extracted and a new one friction-fitted into the main plate from the inside of the movement. This can be done with a jeweling tool or staking set.

Roskopf Watch Cannon Pinions

Roskopf watches are those whose barrel's diameter encroaches over the movement's center. The minute wheel is mounted clutch tight on the barrel cover and it is this that supplies the clutch traction. The cannon pinion, usually brass, must be loose. To restore this clutch action, place the barrel cover and its attached minute wheel on a flat steel stake and a convex solid punch over the pinion. A slight tap renews the traction which is supplied by the burnished overlip of the barrel cover's tube.

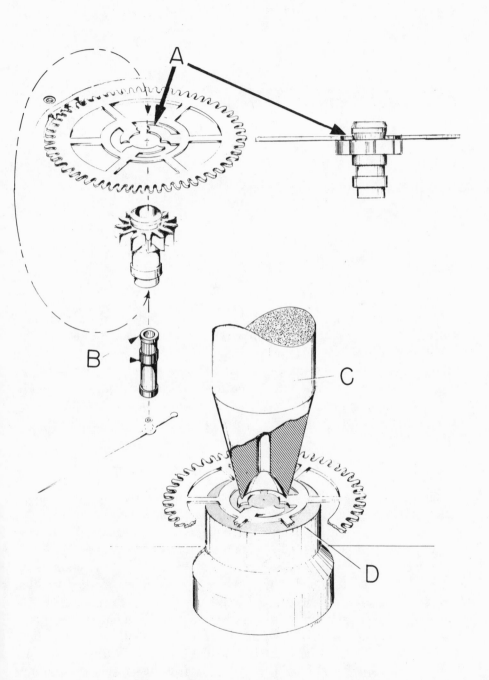

FIG. 12.

V

MAINSPRING REPAIRS

Part I: Repairing Mainspring Endpieces

Frequently a mainspring will break at the end. Because the spring may be an odd size or because a replacement may be difficult to obtain, the end of this spring must be repaired.

There are generally four types of mainspring endpieces: the brace, tongue, "T" end or double brace, and the hole end. Other endpieces are stud end, fish hook end, and combinations of all types of the above-mentioned endpieces, which, however, are seldom found in the popular make of watch.

The Brace End

The brace endpiece is used where space in the barrel is at a premium, such as in small watches. At times, a stronger mainspring is desired and, because a stronger spring is thicker, it requires more space. Therefore, the endpiece must be designed to take up as little space as possible.

The brace is a small stud riveted into a small hole in the mainspring end. To attach a brace end to the spring, a small hole is drilled into the spring by using an old needle or rattail file that has been ground to a three-edged diamond point. This hardened point drills a hole in the spring with a chamfered edge as shown in Figure 1. The "drill" is inserted into a large-size watch screwdriver and is twisted back and forth over the spot where the hole is desired. Drilling is done from the inside surface outward. This type of drill will pierce the spring slowly and the size of the hole may be controlled, giving either a pin-point hole or a large one placed accurately on the spring without any "drifting" to one side.

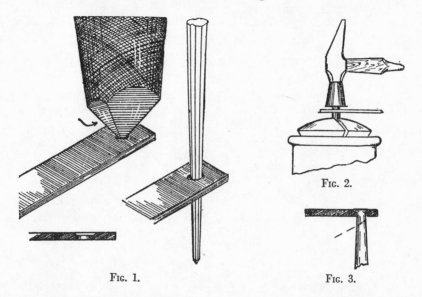

Fig. 1.   Fig. 2.

                          Fig. 3.

When the hole is drilled through or just emerges as a pinhole, a broach can enlarge the hole to a diameter about the thickness of a small pin or the diamond-point drilling can be continued until the required hole size is reached.

A steel, tapered pin is then inserted into the hole so that the thin part of this taper pin emerges from the chamfered side, which becomes the inside of the spring. The tapered pin is cut off close to the spring while the lower part of the taper is held tightly in the pinvise.

Fig. 4.   Fig. 5.   Fig. 6.

The part protruding above the spring is then riveted over by a number of light taps with a small steel hammer. This is shown in Figure 2. The riveted end should then be flush with the spring because it has filled up the chamfered portion of the hole as shown in Figure 3.

At a point shown by the dotted line in Figure 3, the lower part of the taper pin is cut off and filed at a slant. The high part of this pin should be farthest from the spring's end.

The spring is then placed upon a steel bench block or anvil as shown in Figure 4. The hammer then taps the oblique top of the pin so that the pin is hammered back into the form of a hook as indicated in Figure 5.

A thin strip of spring whose ends are filed to a knife edge is then inserted against this hook while the spring is being placed into the barrel. This is shown in Figure 6. The floating endpiece is held against the hook and the opposite end grips the barrel hook. This type does not take up much space in the barrel and is desirable in very small fine watches.

Tongue Ends

The tongue end is most often used in Swiss watches. When there is sufficient spring provided for waste, this endpiece may be produced by

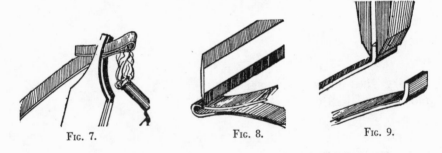

Fig. 7.   Fig. 8.   Fig. 9.

bending over a length of spring while it is held over the flame of an alcohol lamp. As it is being heated, the spring may be pinched together with a pair of flat-nosed pliers as shown in Figure 7. While heating, a heavy pair of tweezers is used to grip the spring so that the heat of the flame is not conducted beyond the end. The heavy tweezers insulates the spring.

Figure 8 shows how a screwhead file is used to cut off the excess length.

Sometimes a tongue end must be produced on a mainspring where no part of the spring's length may be spared. In such a case, the tip of the spring is heated and a flat-tipped or blunt-end tweezer is used to bend over the end by gradual stages of heating, bending, and reheating as shown in Figures 9, 10, 11 and 12.

Frequent reheating and annealing are necessary to prevent the brittle spring from breaking at the bending point. First, the tip of the spring is bent back a little until the resistance offered by the spring is felt to be near the possible cracking point. It should then be heated again, after which it is bent a little more until the resistance serves notice that the tip being bent should be heated once more. This procedure is continued until the bend is complete.

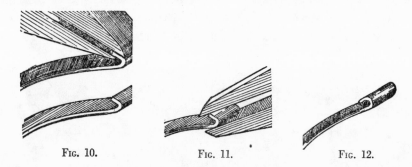

FIG. 10.   FIG. 11.   FIG. 12.

A loose piece of spring (Figure 13A) is provided and beveled at the ends. The beveled piece is slipped into the hook as shown in Figures 13B, C and D. The bent-over lip will pinch this loose endpiece and keep it from falling out. The knife edges assure a firm grip on the barrel hook and prevent slipping inside the barrel.

Riveted Ends

Many mainsprings have endpieces that are fastened with very small rivets. Mainsprings with T ends, double brace and tongue ends, such as those illustrated in Figure 14, are riveted to the end of the mainspring.

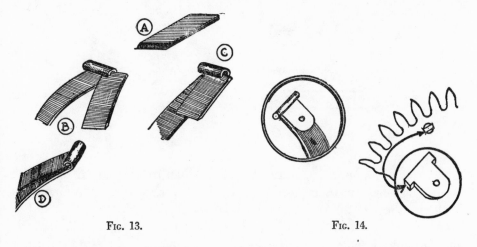

FIG. 13.   FIG. 14.

Sometimes the T end may have to be made. If so, it must accurately fit the hole in the barrel into which it is anchored.

The T end is used in mainsprings to conserve space and also to permit
the entire spring to be wound. The T acts as a pivot and swivels in the hole
near the barrel rim. If a T end with a square brace is used, it will become
jammed into the round hole and bind. When this spring is fully wound,
the brace instead of swinging with the shifting spring may tear or twist
itself off its anchorage. Aside from this, the sharp corners of the "T" brace
may gouge the barrel hole. This is shown in Figure 14.

Fig. 15.   Fig. 16.

Before it is attached to the mainspring, the riveted endpiece is drilled
with a diamond-pointed tool. Figure 16 shows how this is done. The
diamond-pointed tool drills chamfered holes in the endpiece and in the
spring's end. A steel tapered pin is inserted into the spring and out of the
endpiece, chamfer side outward. The pin is cut off close to the brace end,
and the tip is riveted over flush with the spring with a small steel hammer
as shown in Figure 17.

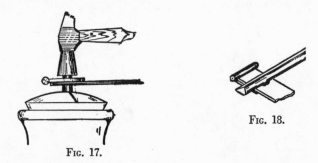

Fig. 17.

Fig. 18.

The remaining portions of the riveted pin that protrude above the spring
or the endpiece are streamlined to the surface of the spring by stoning them
with an oilstone slip as illustrated in Figure 18.

A riveted endpiece should be fastened flush with the spring with no side
overlapping as shown in Figure 15A. If the endpiece is riveted off-center
as in 15B, the wing of the T will not reach into the hole and may result
in a very insecure anchorage. The protruding T will make the over-all

width of the mainspring end wider, and the barrel cover will not stay shut on the barrel.

Should a slightly wider T be used, the brace can be stoned down on its sides to the correct width after it is riveted, as shown by the arrow and dotted lines in Figure 15C.

When the hole in the barrel cover and barrel is a slot or square, the T ends must conform to this shape. However, if the holes are round, the tips of the brace to fit into these holes must be shaped accordingly.

Hole Ends

Some mainsprings come with hole ends. If not, they are easily made by drilling two adjacent holes with the diamond-pointed tool, as shown in

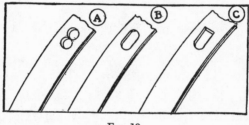

Fig. 19.

Figure 19A. Filing with a fine needle file placed in one of the holes will connect the two holes as shown in Figure 19B.

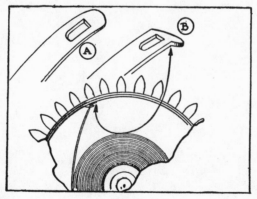

Fig. 20.

The mainspring should be held in a toolmaker's clamp or between the jaws of the pinvise chuck close to the end to permit efficient filing. The holes should be centered and the sides kept straight. A fine square file is used to make the upper end of the hole square, as shown in Figure 19C. Then, the tip of the mainspring is rounded off for appearance's sake and the elimination of any cracks or raw edges, as shown in Figure 20A. The final operation in preparing a hole in a mainspring end is to bend the tip of this spring at the point indicated in Figure 20B. The purpose of this bend is clearly shown. If it is not made, the spring and hole cannot catch the barrel hook and the spring will slip as it is being wound.

SUMMARY

1. A spring may be repaired at its end when sufficient length remains to insure a full day's power to the movement.

2. Where space in the barrel is at a premium, a loose, tongue-end brace may be used.

3. An old needle file ground to a three-edged diamond point will drill chamfered holes accurately in the spring's end.

4. A small steel tapered pin will serve as a rivet after it has been cut to size.

5. When bending over the mainspring, frequent annealing is necessary to prevent the spring from breaking at the bend.

6. Tongue ends should be filed to knife edges so that they remain secure against the barrel hooks.

7. The mainspring should be held in a pliers near its end when heating to keep the heat from traveling back into the spring.

8. T ends must conform to the shape of the holes in the barrel edge.

9. The use of the diamond-pointed tool is recommended because the initial drilling is simple, permitting accurate spotting and observation. It also provides the countersink necessary to permit security of the rivet head.

10. Hole ends should be bent slightly at the tips to aid in gripping the barrel hook.

QUESTIONS

1. Why is it sometimes necessary to repair a mainspring endpiece?
2. When is a loose brace endpiece used in watches?
3. How is the heat prevented from traveling back into the spring while the end is being heated?
4. When is it permissible to make repairs to the mainspring end?
5. What precautions are noted in the case of a T end in the round hole in the barrel?
6. What are the advantages of using the three-cornered diamond-pointed drill?

7. Why should tongue ends have a knife edge?

8. Why is it necessary for hole ends to be bent slightly at the end?

9. Why is it necessary to have the T end centered on the mainspring?

10. How is a hole end in the mainspring prepared?

PART II: HOW TO INSERT A MAINSPRING IN THE BARREL

A mainspring should be inserted into the barrel with the use of a "mainspring winder." If a mainspring is inserted by hand, it is forced into the barrel with the result that the spring will become kinky, the coils irregular, and the spring twisted in the shape of a spiraled cone. Figure 21 shows a spring removed from the barrel after it was inserted by hand. The spring is not flat, but cone-shaped, and the coils do not follow a spiral pattern but

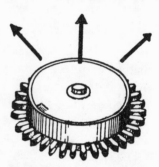

FIG. 21. Distorted mainspring due to inserting it by hand.

FIG. 22. Lines of force of hand-inserted mainspring.

develop with many sharp and sudden corners as though the spiral were made of bending a straight piece of wire in many little bends.

Such a spring is now "set" and has lost most of its resilience and power as well as the ease of motion when it is unwinding. Furthermore, if this spring were used, the lines of force would be divided between upward as well as outward thrusts. This is shown in Figure 22 where the lines of force are shown by the arrows. Instead of giving its power to the edges of the barrel, the spring will exert some of its force upward and, consequently, will rob the watch of needed power to maintain a good balance motion.

Pressure exerted against the barrel cover is cause for the watch to stop, especially with ill-fitting barrel covers.

Mainspring Winders

Mainspring winders cover a wide range of designs and forms, many of which are pictured in Figure 23. However, they all work on the same principle.

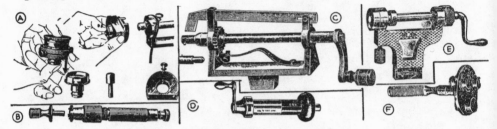

FIG. 23. Various types of mainspring winders.

The spring is first wound on an arbor set into a winder barrel whose diameter is slightly smaller than the barrel for which it is intended. The spring's endpiece is fastened to the anchorage in the *barrel*. The mainspring is then pushed into the *barrel* by a rising platform in the winder.

Winders A, D, E, and F have assortments of winder barrels of fixed diameters graduated in size to accommodate the smallest to the largest mainspring barrels.

The spring is first wound into the winder through an opening in the side of the winder barrel, the spring having been previously attached to the winder arbor that goes through the bottom or over the top of the winder.

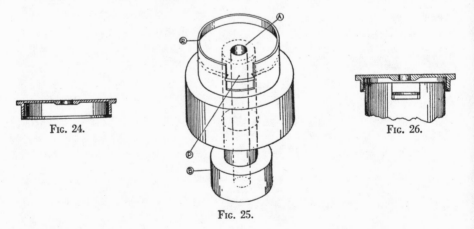

FIG. 24.

FIG. 25.

FIG. 26.

Winder C is used mainly for larger type of mainsprings and barrels. The spring is first wound on the arbor and the finger-brace is pressed on the gathered mainspring to keep it wound. After this, it is slipped into the barrel and unwound into it.

Type "B" has adjustable staves or prongs that may be widened or con-tracted to accommodate a number of barrel sizes.

The winder pictured in Figure 24 is used in this demonstration because it is representative of most of the winders used today. The tongue-end mainspring is also used in the demonstration because it is the most common form of mainspring end and its use with the winders is most easily shown.

Figure 24 shows the winder barrel without its winding arbor. R is the winder barrel rim. A is the hole bearing that guides the winder arbor. B is the push button that raises the platform "P." The platform, in turn, pushes the wound spring into the barrel.

Selecting a Winder

The winder barrel is selected from an assortment of winders so that the one chosen will fit the inside of the mainspring barrel with the tolerances shown in the cross section illustrated in Figure 25. It must fit with a loose tolerance also shown in Figure 26. After the winder is chosen, the main-

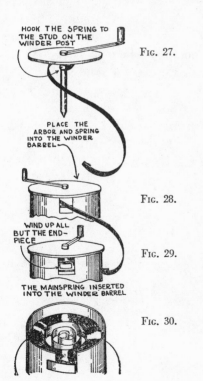

HOOK THE SPRING TO THE STUD ON THE WINDER POST

FIG. 27.

PLACE THE ARBOR AND SPRING INTO THE WINDER BARREL

FIG. 28.

WIND UP ALL BUT THE END-PIECE

FIG. 29.

THE MAINSPRING INSERTED INTO THE WINDER BARREL

FIG. 30.

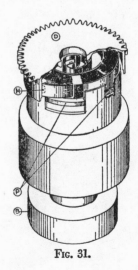

FIG. 31.

spring's inner coil is hooked to the stud on the winder post or arbor as shown in the Figure 27. Care must be exercised that the spring is attached in the correct direction, lest the spring be inserted backwards.

When the spring is attached to the arbor, it is placed into the winder barrel so that the unwound spring is permitted to hang outside the winder through the opening in its side as shown in Figure 28. The spring is then wound clockwise, so that all of the spring but the endpiece is in the winder as shown in Figure 29.

The arbor, cap, and post are then loosened from the inner coil of the mainspring by a slight upward and unwinding motion. The wound spring is pictured in Figure 30. Notice that the tongue-end is outside the winder, not taut inside or on the slot edge.

Inserting the Spring

The wound spring, gathered inside the winder barrel, is now ready to be transferred to the mainspring barrel. As in Figure 26, the winder is placed into the barrel but this time with the mainspring in it. The endpiece is inserted first and the winder barrel is maneuvered so that the endpiece, loose of the winder, is first anchored. Figure 31 shows this. In this view the mainspring is still wound within the winder. But now, the spring is attached by its endpiece to the barrel hook H. The push button B will push up the platform P which, in turn, will raise the wound spring so that it will enter the barrel D, thus being transferred from the winder into the barrel. In this way, the spring remains flat and undistorted.

The spring, inserted into the mainspring barrel, is shown in Figure 32. Here the inner coil's diameter is reduced so that it will grip the barrel's arbor and become hooked on its inner coil hole to the hook in the barrel

Fig. 32. A method of contracting the inner coils to fit the arbor.

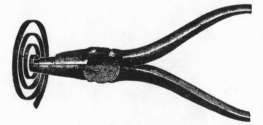

Fig. 33. Mainspring (inner) coiling plier.

arbor. This is done by using a strong pair of tweezers and gently pressing the coil from all sides as shown until the desired fit is secured.

Some watchmakers use a special-made spring coiling pliers to secure this type of coil. Such a pliers is shown in Figure 33.

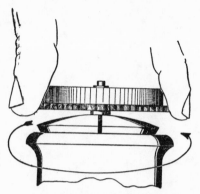

FIG. 34. The security of both ends of the spring is tested by winding up with the arbor in the pinvise and the barrel held in the fingers.

To test the security of the spring's attachment to the barrel arbor, the barrel arbor is placed in the pinvise. The pinvise is then twisted while the barrel is grasped with the fingers. A spring that is securely attached at both ends will permit itself to be fully wound by this method. Of course, the barrel is released gradually after this test is completed. Figure 34 shows the pinvise gripping the barrel arbor and twisting it in the barrel for the above test.

Mainsprings for Self-Winding Watches

In order that self-winding watches do not wind their springs overly tight, the springs' ends have an extra piece attached to them which is about 1/10th the length of the mainspring and about 50% thicker. There is no hook or tongue but these extra pieces act as slip clutches against the smooth inside of the barrel wall. When the springs are almost fully wound, the extra, thicker end slips around the barrel's wall. Until that time, their stiffness will permit the spring to be wound almost to its capacity.

When servicing these, make certain that the brake springs, as they are called, are lubricated with graphite, molybdenum disulphide, or a mixture of this and heavy watch oil. The self-winding mechanism should thus be able to store at least four full turns of the mainspring before slippage takes place. If it slips before this, straighten the brake spring. If it slips too late, it will cause the balance to have overmotion or "rebound" with pos-

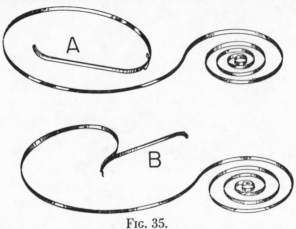

FIG. 35.

sible damage to the roller jewel and pallet fork. In such a case, provide more of a curve to the brake spring. Figure 35 shows two types of brake springs. One is attached to the end of the spring and is wound on the outside of the regular mainspring, while the other is wound on the *inside* of the last coil. When inserting these in the barrel, great care must be exercised not to kink the springs. Figure 34 shows how the tension of the spring may be tested for slippage. Observe the last coil through the slot in the barrel cover. When the barrel is marked *"Sealed Barrel,"* do not take this apart or else you may damage it.

QUESTIONS

1. What is the purpose of a mainspring winder?
2. What damage is possible if a winder is not used?
3. How is a winder barrel chosen?
4. Why must the winder barrel fit into the mainspring barrel?
5. In which direction is the spring wound into the winder barrel?
6. How much of the mainspring is left outside the winder barrel?
7. How is the endpiece secured to the mainspring barrel?
8. How is the spring transferred from winder barrel to the mainspring barrel?
9. How is the inner coil curved to form around the barrel arbor and hook itself to it?
10. How is the spring tested for attachment before the mainspring barrel is placed into the movement?
11. How is the automatic watch prevented from being wound too tightly?
12. What type lubricant should be used in self winding watch mainsprings?

VI

HOW TO MAKE A STEM

A new stem may have to be made when a replacement cannot be secured or the movement has worn so that the stem no longer remains secure in the movement.

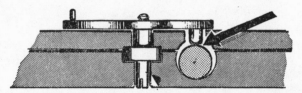

Fig. 1. Stem bearing hole worn, permitting stem to become loosened.

As shown in Figure 1, the stem bearing hole has worn so that the stem has excessive sideshake in this hole, causing the stem to wobble and the winding wheels to strip. Also the setting lever pin is not given a chance to reach completely into the slot in the stem. This results in the stem frequently pulling out. In such a case an oversize stem with parts to fit snugly must be made.

Before going further with the making of a stem, a close study of the stem, its parts and nomenclature should be made. Figure 2 shows a typical stem with accepted nomenclature.

In the making of a stem it is best to proceed with the assumption that the sample is missing. In this way the new stem will be made to fit the movement, winding pinion, clutch wheel and setting lever, custom-built to fit snugly into the worn parts of the movement. No measurements are needed since these are supplied by the movement and parts. These parts become the gauges for the individual sections of the stem.

Metal Used

The material used in making a new stem must be tough enough to do its job of turning the winding and setting mechanism. It must not wear out or cause wear to the parts with which it comes in contact. Furthermore, it must be able to be machined, filed, and finished. An ideal metal for this purpose is drill rod, a high-carbon steel similar to the material of which staffs, knitting needles, or dental burrs are made.

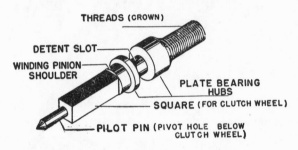

Fig. 2.

To prepare this metal for the making of a stem, it should first be hardened by heating it to a red heat and quickly plunging it into water. Then it is polished so that when it is annealed (softened) the color will show up better. The hardened metal is softened so that it may be machined on the lathe and filed, yet have the toughness specified. This is done by heating the polished, hardened metal over an alcohol lamp flame, moving the rod back and forth over the flame so that the heat is evenly distributed.

In a little while, the color of the polished surface will begin to change. At first it will assume the color of straw. The metal should be manipulated over the flame so that this color is even and affects as much of the metal rod as possible. After the straw color appears, the metal, becoming hotter, will turn to a deeper richer color, resembling that of red wine. As it becomes hotter, this wine color will quickly change to a blue. As the blue reaches the shade of a dark blue watch hand, the rod is again quickly plunged into water so that it will not become any hotter and so that it will retain the blue color. If the stem is made from metal of this color, temper, and toughness, no further tempering will be needed. Tempering the stem after it is made sometimes warps the stem at the setting lever slot.

The piece of metal selected should be of sufficient length and thickness to permit subsequent reductions. It should also be long enough to be held in the lathe chuck.

The next step is to remove the entire setting and winding mechanism. This includes the setting lever, clutch lever, setting bridge and spring, winding pinion and clutch wheel, hour wheel and minute wheel, and setting wheel. It is always a good policy to remove the balance and other delicate parts of the watch to guard against accidents to them.

The blued rod of steel is then placed in the lathe chuck and lathe. It is turned to a smooth, even cylindrical surface so that it will fit into the movement snugly with its end resting against the bottom recess cut for the clutch wheel and resting upon the hole into which enters the pilot pin. Figure 3

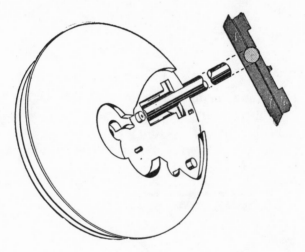

FIG. 3.

shows this. Compare this with Figure 1 where the old stem did not fill the top bearing hole. In Figure 3, the new rod fills the hole completely.

Generally, the thickest part of any stem is the hub. Therefore, if the rest of the metal were turned to the same thickness as shown in Figure 3, it would facilitate the subsequent operations.

The next operation is to cut the pivot of the lower part of this rod. This is to become the pilot pin. The pilot pin must be made so that it will have no sideshake in the pilot pin hole below the clutch wheel recess. The importance of this lies in the fact that it serves as a pivot just as any wheel arbor. Too much sideshake will cause the winding wheels to strip. The pilot pin must also be long enough so that when the stem is subsequently made and is pulled out into the setting position, this pin will not fall out

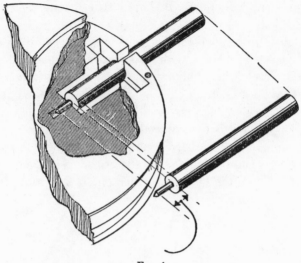

of its pivot hole, enough being left in the hole to brace the stem. Therefore, it is best to cut a longer pin than needed. The result will be that the cylinder cut in the first operation will not be able to go down into the position shown in Figure 4. However by subsequent careful shortening, the longest effectual length is assured as shown in Figure 4.

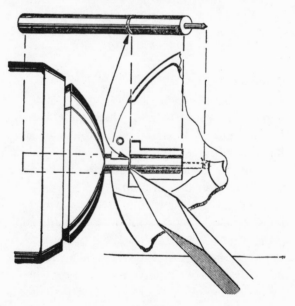

In the making of other parts, components which they accommodate first must be provided for in their logical order. The making of a stem is no

Fig. 6.

exception. The stem must first enter the movement so the bearing hub and pilot pin are first provided. The first wheel that fits onto the stem is the

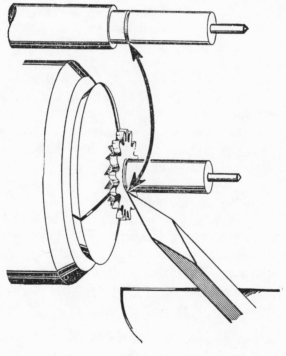

Fig. 7.

winding pinion. Therefore, the stem must be made to accommodate this wheel up to the point in the movement provided for it.

Sometimes the hole in the winding pinion is of the same diameter as the stem hub. If this is the case, the following operation may be passed over.

However more often the hole in the winding pinion is smaller and a cylindrical portion of the stem must be provided for this wheel.

The milled recess in the plate shows where this winding pinion is accommodated. Figure 5 shows the stem as it is made up to now. The stem is still held in the lathe. A sharp graver with its point touching the stem as it emerges into the winding pinion recess will cut a slight marking groove if the lathe is turned slowly. This groove will indicate how far back the winding pinion shoulder must be cut.

The movement is then removed from the stem in the lathe. With the

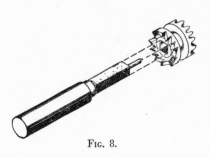

Fig. 8.

winding pinion hole as the gauge, a shoulder is cut to fit snugly into the hole of this pinion. This shoulder must reach from the pilot pin back to the groove just cut as shown in Figure 5. Figure 6 shows how the new shoulder must look after it is cut to accommodate the winding pinion. It must be cut smoothly and fit with very close tolerances. To make this shoulder thinner will permit the winding pinion to wobble on the stem and cause rapid wear to itself and to the clutch wheel.

Following the winding pinion is the clutch wheel. The clutch wheel has a square hole and the surface not used by the winding pinion must now be converted to a square shaft to fit into the clutch wheel.

While the stem is still in the lathe, the winding pinion is placed on the stem but reversed so that its upright ratchet teeth face the lathe chuck and the flat back faces outward as shown in Figure 7. The lathe graver is again brought close to the stem where it emerges from the hole in the winding pinion. While the stem is turned in the lathe, a slight groove is cut as shown in this drawing. The portion of the stem from this new groove to the pilot pin becomes the square shaft for the clutch wheel.

The square shaft must fit into the square hole in the clutch wheel and will then look like Figure 8.

Filing the Square

Filing the square may be done in a number of ways. An elaborate set-up such as the use of a milling attachment may mill or grind the square accurately. However, the square may be filed with comparative accuracy with practiced skill or using a filing fixture.

A filing fixture is an attachment that fits on the lathe bed or into the T rest holder as shown in Figure 9. The filing fixture has two hardened steel rollers parallel to each other and these straddle the projecting stem. A height-regulating nut lowers or raises the yoke into which these rollers are set. This enables the file which rests upon the rollers to work at various heights and file squares or flat surfaces of various thicknesses or distances from the center of the lathe chuck.

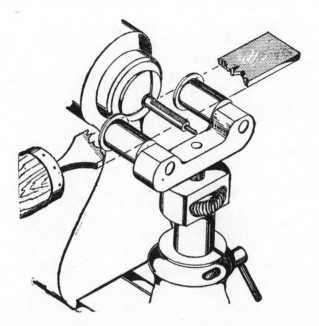

FIG. 9. Filing fixture used to aid in squaring stem.

The rollers are cylindrical rods whose front ends have disc-like shoulders. They keep the file from slipping into the lathe chuck while it is stroked across the rollers or from encroaching onto the winding pinion shoulder.

To set up the stem for filing the square, it is retained in the lathe. The lathe head is then locked in place to keep the lathe from turning or twisting while the square is being filed. In Figure 10, the side-rear view of the spindle is shown. This contains an index plate. This plate most often contains 60 equally spaced holes. Generally, every 15th hole has another indentation above it to mark the four quarters of the lathe head. To lock the lathe for the purpose of squaring a stem, the locking pin, shown in this view, is inserted in the hole at one of the quarter marks indicated by the arrow and in the magnified view.

The filing fixture is then adjusted to a position a little higher than thought to be needed for a trial cut. A fine file in good condition is used. The safe edge should be placed so that it faces the lathe. The file should not cut closer than the groove marked for that purpose. The disc ends of the fixture rollers also serve that purpose.

The thickness of the stem square may be obtained by trying several stems of varying square thicknesses in the clutch wheel. When the square which fits satisfactorily is chosen, two measurements are taken. One is the thickness of the square, Figure 11A, and the other, 11B, is the gauge of the diagonal measurement, from corner to the diagonally opposite corner. The diagonal gauge is the diameter of the shaft which is to become the square. This shaft, shown in Figure 7, from the arrow point toward the pilot pin, should be cut to this new (diagonal) diameter.

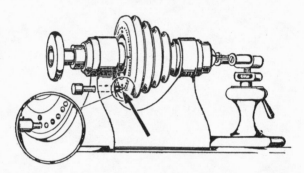

Fig. 10. Index and locking pin inserted in quarter holes.

When the stem is filed, the first surface should appear as the projecting stem in Figure 9.

The file is placed flat across both rollers and pushed forward with strokes that do not bear too heavily on the stem. Pressing down too heavily on the stem may bend it, and a true square will be impossible to achieve. The file should make full strokes and the filing continued until it is felt that the file no longer is cutting the stem since the stem's new flat surface is the same level as the rollers.

The index pin is then pulled out and the headstock twisted so that the locking pin may be inserted in the next quarter hole. The filing is repeated until the four sides are done. The square may not be finished but rather it may have rounded edges as in Figure 12A.

Some new factory-made stems come finished with rounded ends as shown in Figure 12A. However, in a hand-made stem this is not always desirable and the stem should be finished with sharp edges as indicated by the arrow in Figure 12B. This should be indicated from the manner in which the diagonal measurement (Figure 11B) was taken and from a close observ-ance of the clutch wheel hole.

If the clutch wheel does not yet fit, the filing fixture may be lowered just a bit and the four surfaces filed again until the edges meet. The stem *must* not be removed from the lathe during the entire operation of squaring. To see if the clutch wheel fits the square, the headstock may be moved back and the clutch wheel tried on the square for fit. In this way no part of the

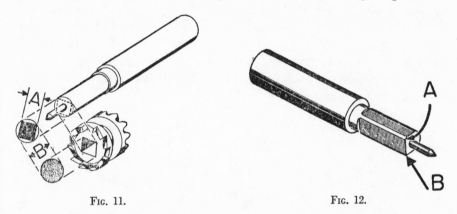

Fig. 11.   Fig. 12.

set-up is disturbed. The clutch wheel should be able to slide along the entire square without binding. If it does not fit, the headstock is brought close to the fixture again. The fixture is then lowered very slightly and the stem filed again. If the fit is very close, it may be stoned with an oil-stone slip.

A well-filed square should permit the clutch wheel to move up and down along the square freely without any twisting sideshake of the stem in the clutch wheel.

Filing the square with a filing fixture assures an even, centered square, each surface of equal width and at right angles to one another.

When the square is to be filed by hand and without a filing fixture, greater care must be exercised. Figure 13 shows how this is done. The stem is in the chuck and the headstock is locked as in Figure 10. The T rest is brought out far in front of the lathe bed and serves as a guide for the file, but the file must not either touch or rest upon it.

To prevent the file from filing the winding pinion shoulder, the left hand grasps the headstock so that the thumb is pointed outward. The thumb is held against the safe edge of the file and guides the file, warding off any pressure that would trespass on the winding pinion shoulder bearing.

When "free-filing" the square, the width of the first filed flat surface becomes the guide for the other three sides. The first flat surface must be filed flat so that each straight edge of this flat surface is perfectly parallel to its opposite edge. If the edge toward the pilot pin seems wider than the edge nearer the chuck, it is an indication that the stem is being filed to a taper. If the opposite ends are wider, the result will be that the clutch wheel will work only in the setting position.

When the first surface has been filed satisfactorily, the next quarter is then attempted. The second surface must be filed to exactly the same width and no wider or narrower than the first. This is a *must*, even if the corners remain rounded as in Figure 12A. The third and last surfaces of the square should then be filed exactly as the first. If one surface were filed wider than another, the result would be a shaft that is rectangular and not square as well as one that is off center.

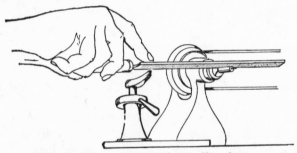

Fig. 13. Filing a square without a filing fixture.

If the clutch wheel does not yet fit, the first surface is filed a little further, keeping in mind that the edges must be kept parallel and true. The other surfaces must then match the new width of the first until the clutch wheel fits. In this way, it will become a square instead of a rectangle and each surface will be equally distant from the center. A square that is off center with the pilot pin will soon wear out all the winding wheels because of the eccentric path which it forces the clutch wheel to take. A stem that is tapered is too thin at one end and permits the stem to skid around in the square hole of the clutch wheel. This is especially true during the setting position. (See Figure 14.)

After the square is filed to satisfaction, it is placed in the movement together with the winding pinion, clutch wheel, clutch lever, and setting wheel. The setting lever is replaced but not tightened. The stem is then tested to see whether it will wind the watch and, when the clutch is shifted, it should easily turn the dial train.

The stem, while still in the movement, is then placed into the lathe again as shown in Figure 15. At a point a little above the top of the winding pinion the sharp graver cuts a little groove (B). This point should correspond with the lower edge of the setting lever pin while in the winding position.

At a point on the stem a little before it enters the movement, a second groove is cut (Figure 15A). This groove marks the lowest point where the threads may be cut.

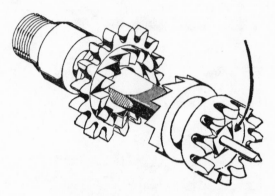

Fig. 14. Square must not be tapered, otherwise clutch will slip in setting position (arrow).

Threading the Stem

The next operation towards completion of the stem is threading. Putting the threads on the stem precedes the cutting of the setting lever slot. The setting lever slot weakens the stem perceptibly and, if this were attempted first, the subsequent strain of threading might break the stem at its weakest point. For the threading operation, a screwplate is used.

The diameter of the threads depends upon the threaded hole in the crown to be used. If the crown is at hand, the thread size may be determined easily by trial fits with some old stems of varying thread sizes. Then the diameter of the proper stem thread is noted and the portion of the stem to be threaded is turned down to this size while the stem is held in the lathe, the chuck grasping the stem at points A and B in Figure 15 with the square well inside the chuck.

The same sample thread that was trial fitted into the crown is now tried in the many holes in the screwplate until the hole in which it fits snugly is found. This hole in the screwplate will be used to thread the stem. Figure 16 shows how the stem should be threaded. The left hand grasps the head-stock as shown, while the right hand holds the screwplate so that the stem may be eased into the selected threading die hole. A little clock oil is used as a lubricant. The actual threading is done very cautiously, as the threads are thin and delicate, stripping easily. The screwplate is twisted a quarter turn on the stem or rather the headstock is twisted a little toward the operator and the screwplate moved a little bit away. This process is then reversed or unscrewed halfway back again. The backward motion is to break

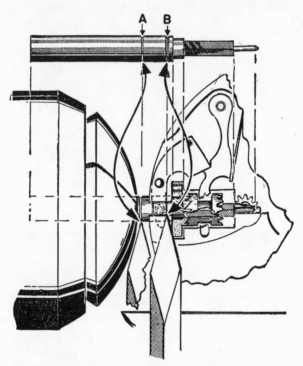

Fig. 15. Groove *B* indicates point where detent neck is later cut. Groove *A* marks lowest threading position.

off any chips that might later clog the threads. The threading is done in these gradual stages a quarter turn or less ahead and a backward one-eighth or more turn to clear the chips.

About every full three or four threads the screwplate is cautiously unscrewed and removed for the purpose of cleaning out the chips. The stem may be cleared by having the lathe reversed and holding a piece of cork against the slowly revolving stem thread.

When the screwplate is returned for the continued operation, it should be lubricated again. The same procedure is followed until the desired length is threaded. Some screwplates, especially older ones and those worn out a bit, half-cut the threads, and partially squeeze the metal into a thread shape similar to knurling. Therefore, it is not uncommon to finish the threading operation and, upon checking, to find that the diameter is thicker than it was before threading. Some watchmakers recognize this and purposely turn the threading shaft .05 mm thinner than required so that they might finish with a correctly gauged thread. In Figure 16, the screwplate is shown cut away, indicating the method of threading.

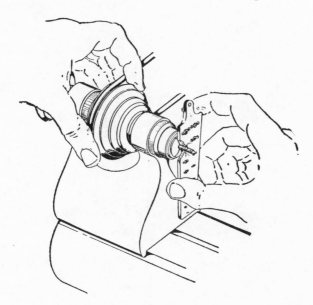

FIG. 16. Threading with a screwplate.

Cutting the Detent Slot

The last operation in making the stem is cutting the slot or groove for the setting lever. This slot must be cut wide enough to permit the setting lever pin to enter without binding. If it is too wide, the slot will not be able to push the setting lever back into the winding position after it has been set. The slot must be deep enough to permit a little freedom between the tip of the setting lever pin and the bottom of the slot. If it is cut too deeply, this will materially weaken the stem. If it is not cut deep enough, the setting lever pin may bind in the slot and may break the setting lever screw, or it may not have enough of a purchase on the stem to hold it and set the dial train in motion.

The slot must be cut so that the setting lever assumes the correct position when it is at rest (in the winding position). If the slot is cut too far up, the setting lever will force the clutch lever and clutch wheel out of engagement with the winding pinion.

Figure 15B shows how the first markings are taken. Here the groove is cut at the point where the setting lever pin contacted the stem. Figure 17A shows the light marking groove cut with the setting lever pin over it. The dotted lines extending down in Figure 17B indicate the proper width of the cut to be taken. Also, the setting lever is shown in the slot properly cut and

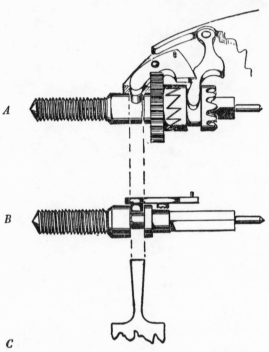

Fig. 17. Showing how wide the detent slot or neck should be cut with cut-off tool C.

the side- and endshakes permitted in this slot are illustrated. Figure 17C shows the tool which is used to cut this slot.

The setting lever slot is cut with a "parting tool" shown in Figure 18. This tool may be made from any high-grade steel shaped to the proportions shown. The dimensions are based upon the necessities and measurements of the setting lever slot. The tool may be ground to size and tempered for cutting use. The better the finish of the cutting edges of this tool, the better and cleaner will be the cut.

The stem is held in the lathe chuck by the upper plate bearing hub and the square sticking out of the chuck. The cutting tool is presented to the

stem as shown in Figure 18 with the T rest as close to the stem as possible
to relieve the strain upon the cutting tool. The slot must have square sides
and corners if a positive setting action is to be obtained. The stem having

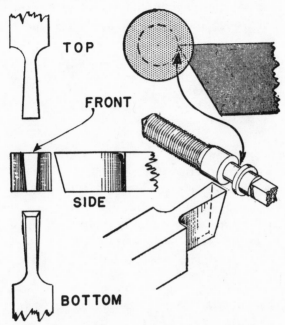

TOP

FRONT

SIDE

BOTTOM

Fig. 18. Specifications for miniature parting tool that is
used to cut the slot.

been previously tempered, no additional hardening or tempering is needed.
The stem may then be tested before the final fitting of the crown.

Summary

1. A worn bearing hole in the movement may necessitate a new custom-
built stem.

2. High-carbon steel, hardened and annealed to a deep blue color,
should be used.

3. The thickness and length of the pilot pin and the fit in the bearing
hubs are most important for the new fit.

4. The square may be cut with a milling attachment or filed with a filing
fixture. An acceptable square may be filed without any fixtures by very
skilled watchmakers.

5. The dividing or index plate generally has 60 holes with 4 quarter holes. Into these holes the index pin is placed to lock the headstock.

6. The safe edge of a fine file is held toward the lathe.

7. The square should not be tapered or left too thin.

8. The square may be finished by stoning with an Arkansas oilstone slip.

9. The threading is done in gradual stages with frequent cleaning of the screwplate and the partially finished threads.

10. The setting lever slot is cut with a parting tool made just for that purpose. The depth and width are dependent upon the diameter of the setting lever pin.

Questions

1. When does it become necessary to make a new stem?

2. What are the two bearing surfaces on the stem?

3. Why must the winding pinion have a good fit on the stem?

4. How long should the pilot pin be made?

5. Of what material should the stem be made and how is it tempered?

6. What is a filing fixture and how is it used?

7. When the square is "free-filed," what method is used to keep the square centered with the pilot pin?

8. How is the square assured and where on the lathe is the index pin?

9. What is the result of a stem with a square that is tapered? Too thin?

10. Describe how the slot is cut and what factors decide the place, width, and depth of the cut and what tools are used for the slotting operation.

VII

FITTING A BALANCE STAFF

The fitting of a balance staff is one of the most important operations in the entire trade of watch repairing.

Fitting a balance staff falls into the following divisions and will be treated in their logical order:

1. Removing a hairspring.
2. Removing rollers (single rollers; combination one-piece "dual" double rollers; two-piece double rollers).
3. Removing the staff:
   a. Riveted type.
   b. Friction type.
4. Selecting and matching a replacement staff with the old sample.
5. Securing the new staff to the balance wheel.

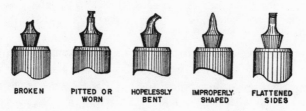

BROKEN   PITTED OR WORN   HOPELESSLY BENT   IMPROPERLY SHAPED   FLATTENED SIDES

Fig. 1. Some reasons why a staff should be replaced.

Balance staffs with broken, worn or pitted pivots, hopelessly bent pivots, badly shaped or flattened pivots should be replaced. (See Figure 1.) The immediate symptoms of a broken balance staff pivot are: watch has

97

stopped; the balance seems loose between the plates; and the balance has excessive shake in all directions. Pivots that are badly bent, worn, or mis-shapen cannot be altered so that the watch will run reliably.

PART I: REMOVING THE HAIRSPRING

Of all the incidents tending to discourage the apprentice, the many mis-haps resulting from the handling of the hairspring are the cause of greatest discouragement. Practically all the damage done to hairsprings may be avoided by treating this delicate part with the respect it deserves. Extreme caution, slow and thoughtful procedure with the use of the proper tools will help to avoid damage.

There are two main methods used to remove the hairspring. Each pro-vides a desired degree of safety and efficiency.

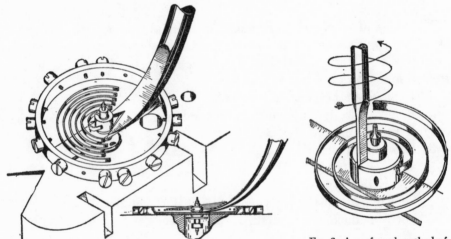

FIG. 2. Prying the hairspring and collet from the staff.

FIG. 3. A preferred method of removing the hairspring.

Figure 2 shows the method used by most watchmakers today. In using this method, a sharpened fork-pronged tool is wedged under the collet of the hairspring resting on the arm of the balance wheel. During this opera-tion, the balance wheel is placed on the utility steel bench block, resting in a hole in the block that fits around the roller table. The forked tool prevents the lifter from slipping and ruining the hairspring. A slight twist upward is all that is needed to remove the hairspring. The use of make-shift but convenient substitutes often results in a badly mangled hair-spring. Figure 2 shows this tool in operation (side view).

Another method of removing the hairspring, used by better watchmakers is shown in Figure 3. In this method a flat-ground needle-like pin is inserted into the split of the collet and twisted upward and off the staff. Twisting the needle spreads the collet open enough to loosen and remove the hairspring without scratching the collet or the balance arm, or distorting the hairspring.

Some watchmakers use both methods, using the needle in the collet slot more often, and, when the collet slot is too small or is closed, using the forked lifter. This simple tool is not generally found on the counters of the material and supply houses; therefore, directions for making it are

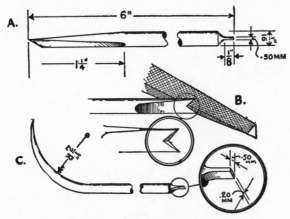

Fig. 4. Details for making the hairspring remover.

given in Figure 4. A piece of drill rod ¹⁄₁₆″ thick and 6″ long is used. An inch and a quarter of the end is filed to a knife-sharp taper (4a). The fork slot is made with a triangular or square needle file held at an angle as shown in (4b). This filing gives the forked appearance as well as sharpening the forked slot. Before curving the forked end, the opposite end is placed into the lathe and cut to the measurement shown. This end should have the appearance of an enlarged balance pivot. The forked end is then curved over the sides of a round file handle. The pivot end is also ground flat to size. Both sides are then tempered and polished with successive applications of emery and crocus cloth.

A simple tempering process for this tool is to heat the ends red hot and quickly plunge into water or oil. This makes it dead hard but brittle. To draw out some of the temper, the ends are polished carefully and then held over the flame of an alcohol lamp, moving the polished end back and forth until the bright steel color changes to a light straw color and then to wine color. As soon as the maroon or wine color appears it should be plunged again into the water or oil. This leaves the steel tough but not brittle. The tool is now ready for use. A little practice with old balances and hairsprings should produce efficiency in this operation. To safeguard the thin needle end, it is inserted into a piece of cork or pith when not in use.

PART II: REMOVING THE ROLLERS

Before attempting to remove the rollers, an examination is made to determine the type of roller used and the method of removal. A single

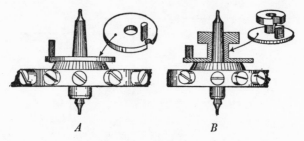

FIG. 5. Single and double rollers.

roller, such as used on the old type of American and Swiss watches, is shown in Figure 5A. The double or dual roller is used on most watches

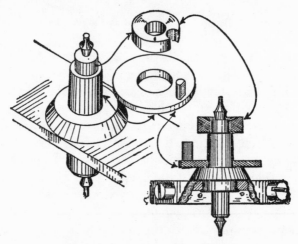

FIG. 6. A two-piece double roller.

today of both American and Swiss type. This is shown in 5B. The two-piece double roller such as used on some Hamilton, Elgin, and Waltham watches is shown in Figure 6. Here the rollers are fitted to their respective shoulders.

There are many types of roller removers. Almost all of these work on the principle that two opposite sharp prongs support the underside of the roller disc or table while a pivot punch, fitting over the pivot and resting on the cone of the staff, is made to push the *staff and balance wheel out of the roller*. An example of this type of roller remover is shown in operation in Figure 7.

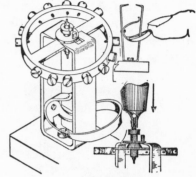

Another method of removing a single or (dual) double roller is shown in Figure 8. The roller is placed into a well-fitting lathe chuck and tightened. The balance is then twisted off the roller. Stubborn rollers may be removed by this method by using an extra lathe head to grasp the hairspring post of the staff and twisting outward. This method is also good for shockproof rollers.

FIG. 7. A typical roller remover and its use.

Removing Rollers from Shock-Proof Watches

Almost every shock proof device uses a roller in which the top of the safety roller is hollowed out to accommodate the bushing for the shock-proofed jewels. Figure 7A shows such a roller. To remove such rollers, the same procedure is followed as when removing the ordinary type of double roller. However, it is a good idea to support the roller, as shown in Figure 7. Extra precaution must be observed with the pivot punch which must drive out the staff through the roller. If a wide punch is used, it will crush the hollowed out, weakened safety roller.

Figure 7B shows how the roller is removed. A punch is used whose tip will safely fit inside the hollow section of the safety roller when the staff

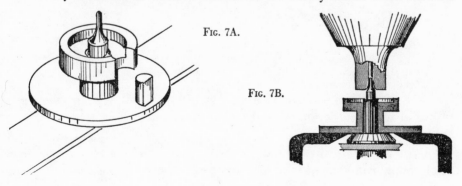

FIG. 7A.

FIG. 7B.

is being ejected, and the tip of the extracting punch must meet the surface of the safety roller's counterbore.

Part III: Removing the Staff from the Balance Wheel

Staffs fall into two main divisions: those that are driven into the balance wheel, friction tight (Figure 9), and those that have undercut shoulders that are riveted over the edge of the hole in the balance wheel to secure them (Figure 10). The method employed to remove the staff, therefore, depends upon the type of staff used in a particular balance. Considerable damage to the balance wheel may be caused by the failure to recognize the type of staff used.

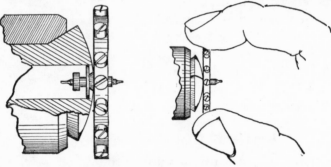

Fig. 8. Twisting the balance off the roller while it is held in the lathe.

To determine the type of staff used, an examination is made of the staff and balance wheel after the hairspring and roller are removed. Generally, the hub of the staff will reveal the information desired. Manufacturers of watches with friction staffs make an effort to notify the watchmakers that

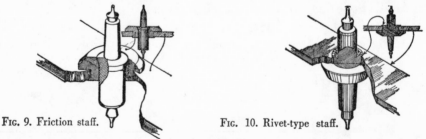

Fig. 9. Friction staff.   Fig. 10. Rivet-type staff.

the balance contains a friction staff. They often will place some (*unusual*) means of identification to warn the watchmakers that the balance hub and the staff are two separate pieces. In the case of the Waltham watches of 13 lignes and upwards, the hub is of blue steel, and the staff of polished steel. Some companies make one roller of bronze and the small (safety) roller of steel in their efforts to inform the watchmaker that these are two separate pieces.

Where the hub is one color and the staff another, the chances are that it is a friction staff. A closer examination with a powerful loupe will reveal the separating lines between the staff and the balance hub. Friction staffs are not generally used in small watches because of the absence of sufficient body around the balance and staff to support the staff rigidly.

Most experienced watchmakers rely upon their knowledge of the make of the watch in knowing whether a friction staff is used. For future reference the watches using friction staffs are listed in table on page 100.

PART IV: REMOVING THE FRICTION STAFF

Friction staffs appear in many styles and patterns or systems of attachment to the balance wheel. Some are driven into hubs which are set into the balance center; others are set or driven directly into a boss on the balance arms; others, like the Swiss MST models or the pinlevers, are simply driven friction tight into the hole in the balance wheel.

In Figure 11, the most commonly used friction staff, the Waltham models, is shown separately and in the process of being removed from the hub in the balance wheel. The Waltham Watch Company recommends the use of special stumps in the staking set to seat the hub. However, since most staking sets do not contain these stumps, a satisfactory method is illustrated here. The balance hub is placed over a hole in the staking set table that is barely smaller in diameter than the

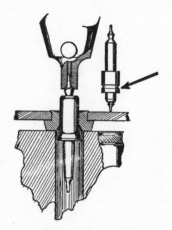

Fig. 11. Driving out a friction staff.

blue-colored hub, yet large enough to permit passage of the friction staff. This hole is then centered with the centering punch and the table locked in place. A driving out punch is then brought over the staff as shown in this illustration and a few light taps will be sufficient to drive out this type of staff. When the staff is broken, there is no need to preserve the pivots. In such a case, the special cross hole punches should not be used since these are delicate and should not be subjected to any unnecessary strain; a small-mouth tapered punch is used.

WATCHES USING FRICTION STAFFS

(American Watches)

Waltham	Hamilton	Illinois
3/0, 12s, 16s	No. 992 16s	16s R.R. Elinvar
	No. 992b 16s	
	No. 915 10s	
	No. 921	

Howard	Hampden	Alden (B.W. Co.)
10s No. 799	L2s Paul Revere	6/0 Central A
12s No. 560	8/0s Mary Jane	6/0s Central B
16s No. 338	11/0s No. 11721	

(Swiss Watches)

Avon Moeris	Civatas	Meyer Studeli (MST)	Admiral (Cyma, Tavannes Tacy)
all models	all models	12s	Os old model
			11 L
			12s old model
Most pinlever watches			16s old model

The Hamilton friction staff is pictured in Figure 11. The Hamilton Watch Company mark their friction staffs with the groove shown by the arrow. This type of friction staff is removed in the same manner as the Waltham type.

The Howard Watch Company used a friction staff similar to the type illustrated in Figure 12. The Howard factory recommended the removal of their friction staffs by placing the balance on a stump or a punch inverted to act as a stump. The hole in this flat stump should fit around the boss or hub in the balance wheel. The driving out punch is brought over this staff and driven out of the balance wheel. The old model Admiral, sometimes called Tacy, Tavannes or Cyma, have friction staffs that resemble the Howard model.

The type of staff illustrated in Figure 13 is used in some of the Central Alden models. Here the staff consists merely of a small, straight, tapered pin pivoted at both ends. This is driven into a one-piece tube, which also serves as the roller and safety roller. Examination of the balance will reveal that the post housing the hairspring is made up of the same piece that goes through a hole in the balance wheel and emerges from the other side as the combination roller table and safety roller. This unit should not be removed.

The staff of such an assembly is removed by placing the balance upon an inverted punch or stump that has a flat face and a hole just large enough

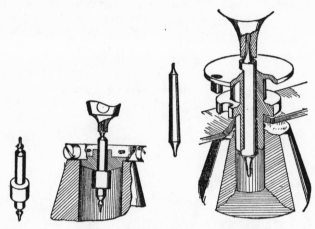

Fig. 12.   Howard type staff.   Fig. 13.

to accommodate the tube surrounding the staff. The illustration indicates a driving out punch forcing the staff out from the direction of its thinnest part towards the thicker end of the slight taper. This will loosen the staff sufficiently to remove it. The punch must not be driven down too far or too hard as this might injure the roller assembly.

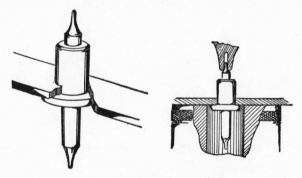

Fig. 14. Simple friction staff.

The cheaper Swiss watches such as the Jagot, MST, Medana, Avon Moeris have staffs such as illustrated in Figure 14. These are removed from the balance wheel simply by selecting a hole in the staking table that fits around the hub of the staff and removing the staff by pushing it out with the punch.

Pinlever, or Roskopf staffs as they are sometimes called, are shown in Figure 15. Figure 15a shows the pinlever staff. The finger-like projection serves as the roller pin in substitution of the more expensive jeweled roller. This finger-like projection fits into the fork of the pallet and receives the impulse from the pallet to send the balance into its vibratory arcs. Figure 15B shows the method of removing this projection from the old staff. It should be removed since new staffs seldom come supplied with it and the old one therefore must be used with the new replacement staff. A cutting tweezer placed under the projection will remove it.

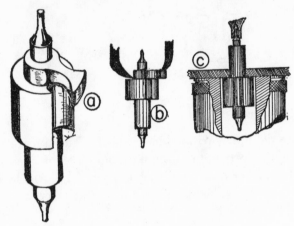

FIG. 15. Roskopf or pin-lever (friction) staffs.

After the projection is removed, the staff is placed in a hole in the staking die in the position shown in 15C. With the balance arm resting on the staking table the punch (shown in this figure), drives out the staff.

PART V: REMOVING A RIVETED STAFF

Subjecting the balance to undue stresses while removing a riveted staff may cause the hole in the balance arms to spread, distorting the balance or splitting it as shown in Figure 16. Since this type of staff has been riveted to the balance, the most logical method of removing the staff without straining the balance would be to cut away the retaining hub of the staff. This will permit the balance to be removed easily, coming off on the roller side of the staff.

To remove a riveted staff by this method, the balance and staff are placed in a lathe chuck that fits over the hairspring shoulder. The chuck is tightened in the lathe and a sharpened, lozenge-shaped graver is pointed against the hub of the balance as shown in Figure 17. Here the back of

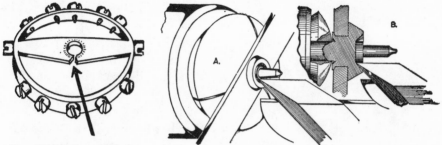

FIG. 16. Balance split due to careless removal of staff.

FIG. 17. Cutting away the balance hub.

the hub is being cut away (Figure 17A). When the graver has cut deeply enough into the hub, it will come to the inside corner of the balance shoulder and the hub. At this moment, the uncut portion of the hub will fall away (Figure 18) as it now has no support from the rest of the staff. The

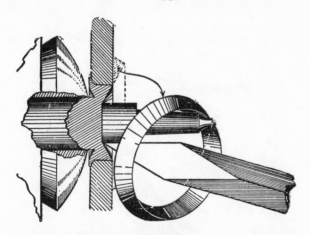

FIG. 18. Balance is loosened when hub is cut off and comes off in form of thin ring.

balance may then be twisted off the staff. However, if it still seems tight upon the remainder of the staff, the balance may be separated by placing it over a stump in the staking set as shown in Figure 19.

Here the hairspring shoulder is fitted closely but not so snug that it may bind. The mouth-tapered punch is fitted over the pivot on the roller post and the cut staff will be ejected with a slight pressure.

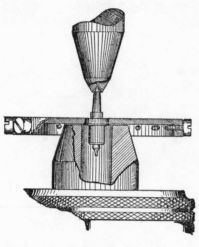

This method of removing riveted staffs is recognized as the safest and will not distort or damage the balance. This facilitates the subsequent operations of truing and poising the balance. A little practice on old 18-size balance wheels will soon give the feeling of skill and the "knack" of doing this job quickly and efficiently.

Sometimes the hub of the balance is rather hard and will resist the edge of the graver. Instead of the graver cutting the hub of the staff away, it will only seem to polish it. This indicates that the metal is hard and that the graver is merely burnishing the staff, thus rendering it even harder and more resistant. In such a case the hub may be annealed (softened) by placing the roller post or the hair-

FIG. 19.

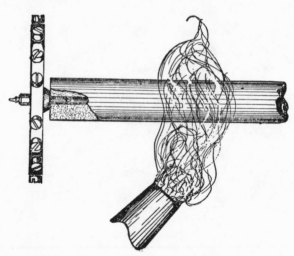

FIG. 20. Annealing a hard-to-cut balance staff hub.

spring shoulder in a close-fitting hollow brass tube 3 or 4 mm in diameter and heating this tube over an alcohol lamp (Figure 20). When the hub of the staff turns a blue color, similar to the blue of a watch hand, it is an indication that the hub is now soft enough to be cut with the resharpened graver.

Sometimes the hardened hub may be cut with special gravers made of carboloy or specially-shaped diamond-tipped tools, specifically made for such an operation.

Using the Staff Extractor

Some watchmakers use a staff extractor to remove riveted staffs. This method, although not as good as cutting off the hub, is better than just knocking out the staff without any support for the balance arms. Because of its wide appeal, instructions on its proper use are given here:

The claims for this tool are that it minimizes the chances of distortion of the balance wheel and prevents the arms from splitting. The principle by which it works is that it serves to clamp the balance tightly upon the staking die (table) while the staff is driven down and out.

A typical staff extractor is shown in Figure 21. To use this tool, the balance is placed on the staking set so that the hub of the staff is fitted closely into a hole in the staking die as shown in Figure 22. The balance arms will then lie flat against the staking table (Figure 23G). The hole is then centered with the centering punch and the staking table is locked. The bottom part of the extractor (23C) is then placed over the top of the balance so that

Fig. 21.

the hole in this part of the tool fits over the hairspring of the staff. This is shown in an "upside-down" view in Figure 23C. The knurled nut (23B) is screwed over the threaded portion of the clamp "C." The extracting punch (23E) is placed into the staking frame, going through B and clamp C, coming to rest upon the pivot of the staff.

The nut B is then turned upward against the staking frame arm A. This action forces the bottom of the clamp C down on the balance arms, pinning it against the staking table D. To permit the extractor to exert vise-like pressure on the balance, the knurled nut B is held firmly between the index finger and the thumb while the arm F is turned in a clockwise direction.

This should be tightened just enough to permit the subsequent loosening of the tool without any strain.

In this condition, the extracting punch is tapped with the hammer, and the staff will be forced out of the balance wheel through the hole in the die H. Since the arms are held rigidly between D and the bottom of C, they cannot bend nor (it is claimed) will the hole spread.

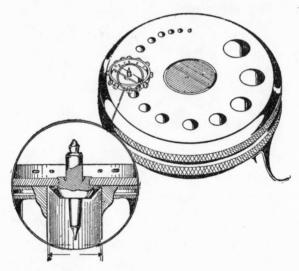

Fig. 22. Selecting the hole in the staking die prior to staff removal.

In some cases the riveted portion of the staff will be sheared off by the sudden pressure and will come off the staff in the shape of a small ring or washer.

The cross hole in C is used to permit examination and to observe if the staff has been extracted. However, a difference in metallic sounds as the punch is being tapped will reveal that the staff is no longer in the balance wheel.

To remove the balance from the staking set, the extractor is first removed by reversing the process of assembly. Examine the hole of the balance and remove all chipped pieces of the riveted part of the staff. Retain the old staff as a model for comparison with the replacement staff.

PART VI: MATCHING A BALANCE STAFF

Many watchmakers are prone to accept a staff for granted, relying upon the word of the supply house or the label on the envelope. They may make a preliminary fit of the balance and roller. However, after the staff is staked to the balance wheel, the errors become apparent.

Although it may be possible to obtain genuine staffs for many types of watches, these do not always fit. This may be due to the hole in the balance wheel becoming enlarged through previous careless handling and removal of staffs. The jewels may have been replaced or the roller table may fit loosely because the hole in the roller has been stretched by forcing it upon

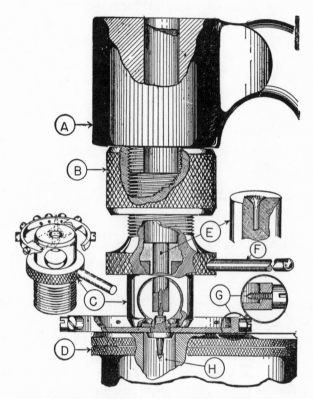

Fig. 23. Cut-away view of staff remover in use.

a staff that was oversize on that part. For these reasons it becomes necessary to make a comparison between the various parts of the new staff with the dimensions and characteristics of the old one.

Since frequent reference will be made to certain parts of the staff, a complete knowledge of the names of these parts is desirable. Figure 24 shows a typical balance staff of the rivet type with the nomenclature clearly indicated. Figure 25 shows this staff with the dimensional lines indicated and lettered in the sequence of their comparative importance. A is the height over-all, or the distance between the cap jewels; B is the thickness of the shoulder that fits into the balance wheel; C is the thickness of the roller post measured at the base of the post; D is the thickness of the hairspring shoulder; E is the distance from the balance seat to the tip of the lower pivot.

The height over-all is the most important measurement of the staff. This decides whether the staff will equal the distance between the lower cap jewel (endstone) and the upper balance bridge cap jewel. A staff that is shorter than the height of the sample will not receive the support of both balance hole jewels and the watch will not run. Staffs that are

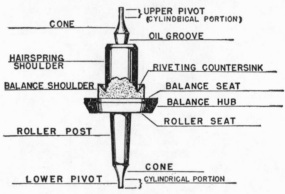

FIG. 24. Nomenclature of typical rivet type of staff.

slightly higher may be turned down to size on the lathe with skillful manipulation.

When measuring a broken sample staff with a broken pivot, 0.25 mm may be added to the total height for each broken pivot if it is a wrist-watch staff. Pocket-watch staffs have slightly longer pivots and 0.30 mm may be added for each broken pivot. For example, let us assume that both pivots on a 10½ ligne man's wrist-watch are broken. Measuring the staff from pivot stump to the opposite one, the micrometer reading is 3.07 mm. Adding 0.25 mm for each of the two missing pivots gives 3.57 as the total height. This will serve as a close estimate of the measurements for the new staff.

Measurement E is very important because this height decides in what plane the balance wheel will move. If the distance E is too great, the balance will be too high and scrape or touch the center wheel or the bottom of the balance wheel bridge. It might also cause the balance arms to scrape or catch upon the regulator key.

Figure 26 shows three staffs, all the same height with thickness of all parts alike. Yet the staffs B and C are not suited to the sample A. Staff C, if fitted, will cause the balance to scrape the pallet bridge and the roller table to touch the pallet fork. The hairspring will not lie flat, being pushed upward. Staff B, if fitted, will cause the opposite effect. Here the balance will be in too high a plane and scrape the center wheel or the regulator key. The roller may also be too high, failing to engage the pallet fork. It can readily be seen why measurement E is important.

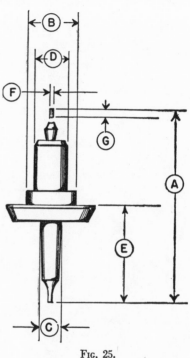

Fig. 25.

Gauging a staff can be done with a micrometer or some reliable gauge. In order to guard against the loss of the sample, measurements of the staff should be taken immediately after the staff is removed from the balance. It is suggested that a freehand sketch of the staff be made and the measurements noted on the left side of the sketched staff. The measurements of the replacement staff may be noted on the right side so that a quick comparison of the values of each are noted. Figure 27 illustrates this procedure.

In the absence of a micrometer or gauge, the balance wheel hole, roller table, and hairspring collet may serve as the gauges themselves. After lining up the replacement staff with the sample as in Figure 26 and if the heights A and E seem acceptable, the new staff is placed in the balance hole. It should fit closely and snugly. If it is too loose, the staff will require excessive spreading with the result that the balance will not be secured firmly to the staff and the punch may harm the balance. A fair test for fitness used by many watchmakers is to place the staff in the balance with the tweezers. The staff should fit closely enough so that when the balance wheel is turned upside-down, the staff will not fall out of the balance hole. A staff that fits tightly is acceptable. A staff that is too thick for the balance hole may be turned down on the lathe. The balance hole should never be enlarged to accommodate such a staff, remembering the rule that no part of the watch should be altered to fit a strange piece of material—rather fit the material to the watch.

Another factor to be considered is that the balance wheel shoulder must

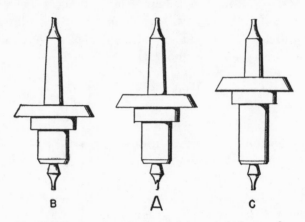

Fig. 26. Three staffs with same height and thicknesses but different in characteristics.

be of sufficient height to permit enough metal to be riveted over the edge of the hole in the balance wheel. In Figure 28B, the riveting shoulder is too low. Here the punch will engage the balance rather than the staff and spoil

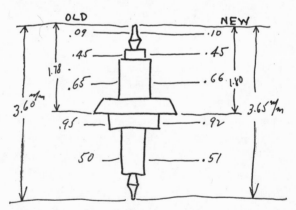

Fig. 27. A rough sketch should be made of the sample and its gauges noted.

the balance wheel. The correct proportion in this part is shown in 28A.

To test the thickness of the roller post, place the roller table over the roller post as in Figure 29. A well-fitting staff will permit the roller table to be pushed down with the tweezers about three-quarters of the roller post before being staked down to the balance hub. The extra quarter length is needed to permit the roller table to be driven downward on the tapered .post to become firmly secured.

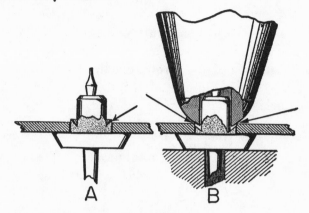

FIG. 28. *A*: Correct height of balance seat. *B*: Result of shallow balance seat.

If the roller post is so thick that it will allow the roller table to go on but part of the way as in Figure 29B, it may be turned or ground down so that the preliminary fitting of the roller is more like in 29A. Driving the roller downward on such a thick roller post will cause the roller table

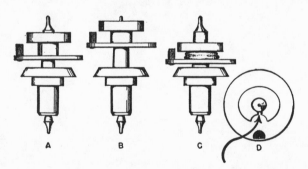

FIG. 29. *A*: Correct pre-fit for roller. *B*: Roller post too thick. *C* & *D*: Results of forcing a roller on an over-thick roller post.

to become crushed or cracked as in 29C. In such a case the roller table is useless.

Rollers that go down further than three-quarters in the preliminary fit will not remain firm on the roller post and will work loose later.

To test the thickness of the hairspring post, the hairspring collet is slipped onto the hairspring shoulder of the staff while it rests on the bench block. If the hairspring collet slips on this shoulder with just a little pressure, it is acceptable. A staff with too thick a hairspring shoulder or one too thin will only cause trouble or mishap to the hairspring. A staff with a hairspring shoulder a few hundredths of a millimeter thicker than the original staff will cause the hairspring collet to spread or crack as well as to distort the inner coils of the spring. A thin, hairspring shoulder will not permit a firm grip, and efforts to close the collet so that it grasps the staff will only cause the collet to crack or become misshapen.

Testing the Staff for Pivot Thickness

A simple method of testing the pivot thickness of the new staff and its suitability to the balance jewels is to place the pivot into the back end of

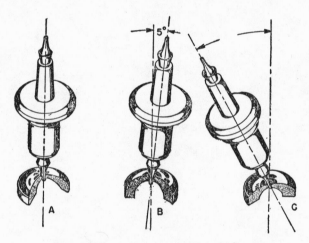

Fig. 30. Testing for correct pivot thickness.

the hole of the balance jewels and to notice its inclination from the upright as shown in Figure 30. Both pivots should be tested in their respective jewel holes. For this purpose the cap jewels should be removed. A pivot that is of the proper thickness, neither too tight nor too loose in the jewel, would have just enough freedom inside the jewel holes to permit the staff to incline from the upright about 5° when permitted to stand alone (Figure 30B). A pivot that is too thick will bind in the jewel and not incline at all, standing upright as in Figure 30A. A balance staff pivot that is too thin will have too much sideshake in the balance jewel holes and by test will incline excessively in the jewel holes (Figure 30C). Such a pivot may cause the watch to stop when it is held in the pendant position.

Pivots that are too thick may be ground and polished to correct size and fit. All corrections and adjustments to the new staff should be made before staking the staff to the balance. This is advisable because then the staff may be easily handled and the balance is not likely to be damaged. Furthermore, should adjustments to the staff be unsatisfactory, the trouble of removing them from the balance is avoided.

Figure 31 shows the staff in the sectional horizontal plane. Here the height of the staff is indicated between the cap jewels with the component parts of the balance assembly and pallet. A well-fitted staff should provide for a slight amount of endshake, indicated by A in Figure 31. The lower part of the balance should have sufficient clearance above the pallet bridge,

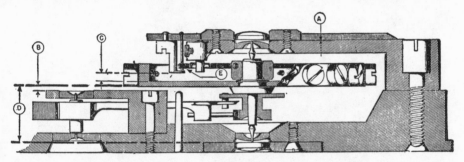

FIG. 31. Sectional view of staff and its proportions in relation to other parts in the watch.

indicated by B. It should also have clearance above this point so that the regulator key does not come into contact with the upper part of the balance wheel arms shown by the lines C. D is the height of the staff measured from the lower cap jewel or the tip of the lower pivot to the balance wheel seat. E, the hairspring, has its collet resting upon the staff and balance and remains in a level plane.

Should the movement be turned over, the endshake (A) is kept to such minor proportions that the distances B, C, and D are not disturbed enough to cause the regulator key to scrape the balance arms or the balance to scrape an overlapping center wheel.

PART VII: REPLACING THE STAFF

After the new staff has been chosen and "matched," the old staff is retained and used as a "stand-in" for the various set-ups prior to inserting the replacement staff. In this way, the chances of damaging or losing the good staff are greatly reduced.

Friction Staffs

We will first take up the task of replacing a friction staff of the Waltham type. In this type of staff, a stump of the staking set used to support the balance during this operation is first chosen. In this job the staff is driven into the hub of the balance wheel. To choose the proper stump, the old staff is grasped by the tweezers at the roller post. The hairspring post of this staff is inserted into the hole of a flat-faced hollow stump that fits this

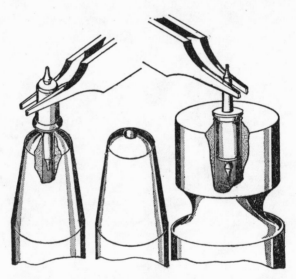

Fig. 32. Selecting punch and stump for driving in friction staff.

shoulder closest without binding. This is shown in Figure 32. The punch used to drive the staff into the balance hub is also shown in this view. This punch is a round-faced hollow punch that is selected to fit over the roller post with a minimum of tolerance. The reason for its use is that it will drive the staff snugly into the balance hub as far as possible.

To prepare the stump for this operation, it is placed into the hole of the staking die used expressly to hold all stumps. Then it is centered with the centering punch and locked in place. The staff is placed into the balance with the tweezers, and the part of the hairspring shoulder emerging from the balance is placed into the stump. The round-bottom punch selected is then placed over the roller post, and a few light taps should drive this staff snugly into the balance hub. This is shown in Figure 33. Here the round-bottom punch has reached into the slight recess in this hub to permit the shoulder of this friction staff to be driven fast and slightly below the level of the top of the hub, providing an extra degree of security.

Replacing a Friction Staff—Hamilton Type

The operation in selecting and centering the stump is the same as that for the Waltham type. However, the punch chosen may be a flat, hollow punch fitting over the shoulder that will hold the safety roller. The requirements of a good fit are the same as for the Waltham types.

Figure 34 shows this set-up in the process of being driven downward until it has reached the balance hub and has come to rest upon it.

The Howard friction staff is inserted in much the same manner as described above for the Hamilton model. The exception here is that the balance, hub, and hollow hairspring post are all one piece. Therefore, the stump used must fit around the hollow section that is used to house the hairspring collet. The

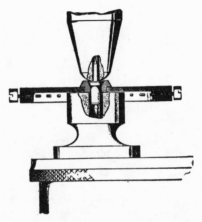

Fig. 33. Driving in the friction staff (Waltham type).

punches and operations follow the same pattern as for the Hamilton type.

The Central A and B model staffs are inserted into the balances with the balance resting upon the stump fitting the portion of the staff that will support the hairspring collet. Since these staffs are plain tapered, pivoted needles, the only edge upon which a punch may rest is the cones of the pivots. Therefore, the driving punch used is a cross-holed pivot punch. These punches are delicate and must be chosen and used with care. The punch chosen must fit the pivots freely, and the holes in which the pivots rest must be clean and clear.

The plane in which the balance rotates depends upon the depth to which the staff is driven. If the staff is driven in too far, the inverted type roller-balance hub will rest upon the bottom plate, interfering with the free motion of the balance as well as causing the balance to scrape the pallet bridge. Failure to drive the staff to its proper depth may cause the balance to lie in a plane too high. In such a case, the hairspring may scrape the balance bridge. The balance wheel and the pallet fork guard finger may come into contact with the roller jewel. Figure 35 shows how this staff is driven into the balance unit.

Figure 36 shows how a simple friction staff of the Avon type and some pin-lever staffs are inserted. Here only the hole in the balance arms provides support for the staff. Therefore, the staff chosen must be a very tight

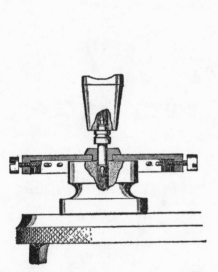

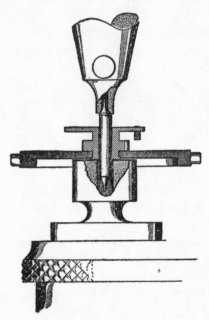

FIG. 34. Driving in friction staff, Hamilton type.

FIG. 35. Driving in staff, Alden type.

fit if truing of the balance is to be attempted without subsequent loosening of the staff. The punch fits over the roller post and the hairspring post of the staff fitting into the stump.

Fitting Pin-Lever or Roskopf Staffs

The set-up and fitting precautions of fitting pin-lever and Roskopf staffs are much the same as those used for the operations described for the Avon type staffs. The balance rests upon the stump into which fits the hairspring post of the staff. A flat hollow punch fits over the finger projection shoulder and rests upon the hub of the staff used as the safety roller. This staff, like other friction staffs, is driven down until it is flush against the balance. Figure 37 shows the staff as it is being driven into the balance.

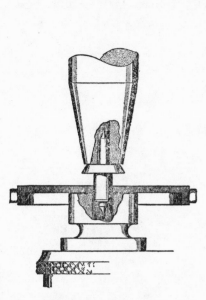

FIG. 36. Inserting friction staff, Swiss type.

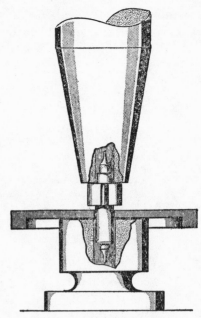

FIG. 37. Driving in Roskopf staff.

PART VIII: INSERTING A RIVETED STAFF

Replacing a riveted staff calls for great precaution and care in each operation from the selection of the staking die holes to the riveting operations. The selection of the proper punches has much to do with the successful completion of this job. As in the fitting of all balance staffs, the old staff is used to aid in the selection of the proper punches and in the alignment of the selected hole in the staking die.

The first step is to choose a hole in the staking die that fits closest around the roller post without binding. (Figure 38.) This hole is then centered with the centering punch and locked. (Figure 39.) Next, a flat-faced hollow punch that fits over the hairspring shoulder of the staff without binding and rests upon the balance shoulder is selected as shown in Figure 40. A punch that will spread the riveting undercut in the balance shoulder of the staff is chosen next. This will be a round-bottom, hollow punch that has the same dimension as the flat, hollow punch. This round-bottomed punch, however, will fit into the recess of the balance shoulder, subsequently spreading it when tapped with a hammer. This punch is shown in Figure 41.

When the staff fits tightly over the balance shoulder, a punch is used especially to drive the balance snugly onto the balance shoulder and against the balance hub of the staff, prior to riveting. This punch is a flat, hollow

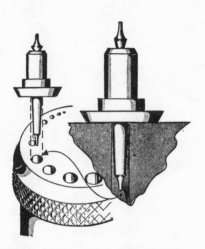

Fig. 38. Selecting hole in staking die prior to rivet-type staff fitting.

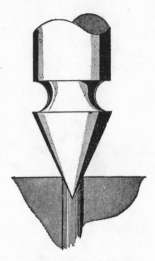

Fig. 39. Centering die hole with centering punch.

punch that fits over the balance shoulder of the staff with little tolerance. The balance is then placed over the staff and this punch, shown in Figure

Fig. 40. Selecting flat faced hollow punch.

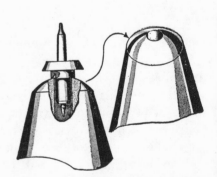

Fig. 41. Choosing a riveting punch.

42, drives the balance firmly onto the staff, readying it for the subsequent operations.

With the balance and staff placed into the staking set, the actual opera·

.ion of securing the staff to the balance may begin. The riveting punch, used to peen the riveting metal over the edge of the balance hole, is placed

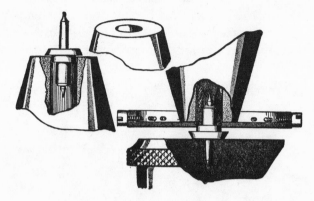

Fig. 42. Selecting punch to drive balance onto tight-fitting staff prior to riveting.

into the staking set and permitted to rest over the staff so that its ball-like bottom fits into the recess of the staff. The top of the punch is grasped be-tween the thumb and index finger of one hand while the light hammer taps the punch. During this tapping of the punch by the hammer, the thumb and finger twist the punch with every tap of the hammer. This permits the metal all around the staff to be riveted with equal force. The effect of the riveting punch upon the staff is shown in Figure 43. The tapping of the hammer should be conservative. Hammering too heavily will only spread the staff excessively with damage to the balance.

A test to determine if the balance has become secured to the staff is shown in Figure 44. Here the thumb of one hand

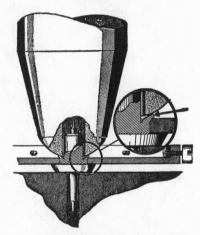

Fig. 43. Spreading over the riveting countersink on the staff.

presses hard down upon the riveting punch. A strong pair of tweezers grasps the arm of the balance, attempting to turn it. Since the punch touches only the staff, the pressure is not upon the balance. If the balance can be twisted on the staff, it is a true indication that the balance is not sufficiently secured to the staff. Therefore, the riveting operation must be resumed until the balance is sufficiently secured to the staff and the tweezers fail to turn the balance upon the staff.

The finishing punch is then used to flatten the riveted lip. This is shown in Figure 45. A few light taps of the hammer upon this punch is all that is needed to make the riveted portion appear as shown by the arrow in this view. This operation also permits the hairspring collet to rest closer to the balance.

After this operation, the balance is removed from the staking set and placed into the movement. It is then tested for height and freedom in all

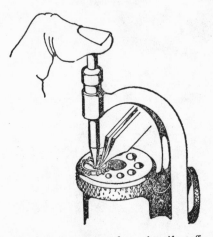

FIG. 44. Testing to determine if staff has been sufficiently riveted.

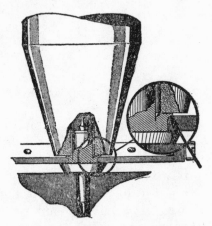

FIG. 45. Flattening or "finishing" the riveted ledge on the staff.

positions while it is spun with the aid of a soft camel's-hair brush. Notice if there are correct endshake, pivot freedom, interference with the pallet bridge and any protruding parts near it. The rollers are then placed on the staff.

Top and Side-Grooved Staffs

Before replacing riveted staffs, examine the riveting groove. If this is very deep, the riveting should be done with a flat-faced, hollow punch alone. The Hamilton Watch Company calls these the "Top Grooved Staff." The claims for these staffs are that they are easily riveted and that subsequent removal in the case of a broken pivot is easily accomplished without the use of a lathe or staff extractor.

The Elgin Watch Company has a side-grooved staff which also requires only a flat-faced, hollow punch to fasten it to the balance arm. This staff is easily recognized by the slight grooved indentation in the side of the balance (riveting) shoulder. Like the Hamilton Watch Company, the claims for this staff are the easy removal of the staff, when broken, without lathe work or staff remover. This is also explained in the Elgin technical data sheets describing this staff.

PART IX: REPLACING ROLLER TABLES

Replacing a single roller calls for the use of special slotted stumps to support the roller without breaking the roller jewel pin while the balance is driven into the roller.

Such a stump is shown in Figure 46. The roller is placed upon the balance staff roller post as shown in Figure 47. Notice that the jewel is

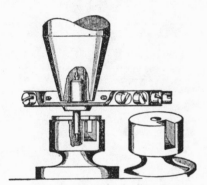

FIG. 46. Use of stump for single rollers.

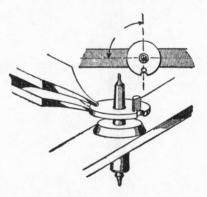

FIG. 47. Placing the single roller in position prior to driving it in place.

placed at right angles to the balance arms. This is true of all rollers except where otherwise recommended for specific models by the manufacturer. The stump chosen must fit over the roller post with close tolerances. The slot must be sufficiently wide and deep enough and cut close enough to the hole to permit freedom of the jewel pin without binding or subsequent cracking of this jewel. The balance and roller are then placed over the stump. The finishing punch used and shown in Figure 45 is used to drive the balance and staff snugly into the roller.

Replacing a Two-Piece Roller

In replacing a two-piece roller, such as used on Hamilton watches, the balance is placed on a flat, hollow stump fitting around the hairspring shoulder of the staff. The balance is placed, bottom up, in this stump. A slotted hollow punch, fitting both the large, jeweled roller and the roller post of the staff, is selected. The punch should have all the specification of the stump shown in Figure 46. Special care is taken that the roller post does not bind in the hole in this punch. The roller is then driven down flush to the balance hub of the staff.

To place the small safety roller in position, a flat, hollow punch whose hole fits closely over the safety roller shoulder of this staff is then used. The safety roller is carefully placed over the post with the tweezers so that the "passing" crescent is directly in line with the roller jewel. The punch is then used to drive this roller down into position. This is shown in Figure

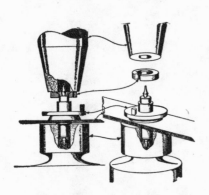

Fig. 48. How a two-piece double roller is replaced.

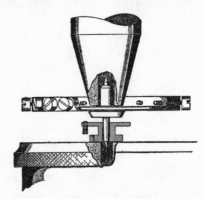

Fig. 49. A method of replacing a double, one-piece roller.

48. Care must be used that the roller does not require excessive driving to bring it flush with the shoulder. These rollers are brittle and any undue force may crack these rollers at the crescent. When replacing the rollers, it is best to place the roller on the post so that the jewel is on the side that it was originally, so that the delicate poise of the balance is not overly disturbed.

Replacing a Double (Dual) One-Piece Roller

Replacing a double roller is accomplished by setting up the staking set as in Figure 38 with the roller post fitting closely into the hole in the die. The roller is placed on the roller post so that it will not fall off when turned over. The finishing punch shown in Figure 45 is used to drive the roller table down to the balance hub. This is shown in Figure 49.

The tapping of the punch in driving down the rollers should be done with care and observation lest the roller be crushed. While the roller is

being driven down, the sight is trained upon the space between the roller
and the balance hub so that the tapping on the punch may cease when the
roller is down in place. A difference in metallic sounds will also tell when
this occurs.

When all rollers are in place, the balance is placed into the watch and
tested in all positions for impulse, freedom of motion, and all points

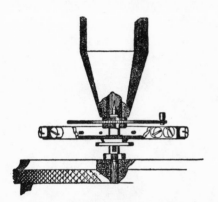

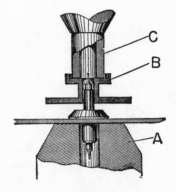

FIG. 50. Replacing the hairspring   FIG. 51.

stressed in Figure 31. After all these have been proven satisfactory, the
balance is trued and poised.

The hairspring is then placed upon the balance staff so that the stud is
in line with a point on the balance rim which will insure an "in beat" position. To place the hairspring securely on the watch, the proper and safe
method is to place the balance and rollers on the staking set exactly as
described in the paragraph on replacing a dual roller. The same punch
and staking hole are used. The forefinger pressing upon the punch is all
the pressure that should be required to force the hairspring collet down
upon the hairspring shoulder of the staff and snug against the balance
arms, as shown in Figure 50.

Replacing Rollers on Shock Proof Staffs

Rollers which are used with shock proof staffs and with shock resistant
devices, such as Incabloc, Kif, Monorex, Ruby-Shock and others, have
counterbored tops on the safety roller to accommodate clearance for the
bushings for these devices. In replacing a roller to such staffs, special
punches and methods must be used or else the weak safety roller will be
easily squashed. The proper method of replacing these rollers is shown
in Figure 51. Place the balance on a staking stump A, whose hole allows a

slight clearance to the hairspring shoulder of the staked staff. Next place the roller B, upon the staff loosely. Then select the special roller punch C, whose hole will provide clearance for the roller post of the staff and whose special cylindrical section is thin enough to enable it to fit easily inside the counterbore atop the safety roller B.

It can be seen from this Figure that if an ordinary flat, hollow punch is used, it will be astride the top of the safety roller and any pressure exerted by such an incorrect punch will cause the rim of the safety roller to collapse. The special punches needed are purchased in sets to fit any size shock proof rollers.

QUESTIONS

1. When is it necessary to replace a balance staff?
2. What are the symptoms of a broken balance staff?
3. Why shouldn't a screwdriver be used to remove hairsprings?
4. How is the two-piece roller distinguished from the dual or double roller?
5. What are the two types of staffs?
6. How is the friction staff held in the balance?
7. How is the riveted staff held secure in the balance?
8. How is the friction staff removed from the balance?
9. What is the function of a cross-hole punch?
10. What is a Roskopf staff?
11. What danger is present when a riveted staff is driven out?
12. What is the best method of removing a riveted staff?
13. How may a balance hub be softened without danger to the balance?
14. Into what hole in the staking die must the balance staff hub fit when using the staff extractor?
15. Make a sketch of the rivet type staff and name all its parts.
16. How is the pivot tested for proper fit in the hole jewel?
17. How far down should the roller go on the trial fit?
18. Why is the height of the roller post of the staff important?
19. Why is the old staff retained even after matching?
20. How should the punches fit over their respective shoulders of the staff?
21. What is the test to determine if the balance has been firmly secured to the staff?
22. How should a hollow top roller be replaced on a shock proof staff?

VIII

HOW TO MAKE A BALANCE STAFF

When a duplicate balance staff is unavailable and the nearest "match" requires too many adjustments, it is better to start from the raw material and make a new staff.

One often hears that this skill is unnecessary, but it is hardly necessary to point out that in everyday watch repairing a competent watchmaker makes many different alterations in the various staff jobs he encounters and, actually, the sum total of these represents the equivalent of the work involved in making the complete staff. One who can make a staff certainly is able to alter a ready-made staff to fit.

This chapter will deal only with the making of the staff. Finishing and polishing have been covered in the chapter dealing with "Adjustments to the Balance Staff."

When the old staff is available and has served well up to the time of its breakage, it may be used as a sample. To guard against its being lost, a sketch is made and the measurements listed, similar to the manner shown in Figure 1.

It will be noticed in this illustration that many of these dimensions may be secured by mathematical deduction. For instance, the height of the hairspring post to the tip of the pivot (C) is obtained by deducting height (D, E, B) from height (A). The length of the roller post (B) is obtained by subtracting heights (E, C, D) from (A). This system may be continued until all individual heights or gauges are secured. A review of the chapter, "How to Match a Balance Staff," will also help in understanding how the measurements are obtained.

Sometimes the old staff has been lost or its characteristics and measurements have been found faulty. In such a case, the new staff will have each part made "custom built" to fit according to its needs and available space.

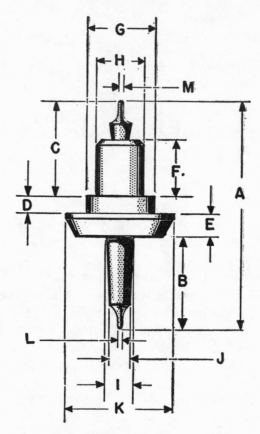

Fig. 1. Important dimensions of a typical balance staff.

The thickness of the pivots and the shoulders of the staff are obtained from the parts into which they fit or parts which fit upon it. The heights of the shoulders and hubs are obtained from the movement. A millimeter micrometer may be used to aid in obtaining these measurements.

Height Over-All

The complete length of the staff, from pivot tip to the opposite pivot end, is called the "height over-all" (A, Figure 1). This is obtained by using the micrometer to gauge the thickness of the movement at a point directly over the upper and lower cap jewels as shown in Figure 2. From this measurement (D, Figure 2) subtract the thickness of the upper cap jewel (B) and the thickness of the lower cap jewel (C). The result will be the height of the new staff (A, Figures 1 and 2).

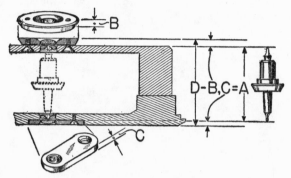

FIG. 2. Determining the over-all height of the staff.

The next measurement to be obtained is the height of the roller post (B, Figure 1). This includes the lower pivot. The height of this post is obtained by gauging the movement from the top of the pallet bridge to the bottom of the lower plate (A, Figure 3), as shown. Subtracting the thickness of the lower cap jewel (C, Figure 3) from this gauge *will* give the height of the roller post measured from the tip of the lower pivot to the roller seat on the balance hub.

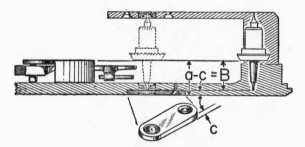

FIG. 3. How to secure the height of the roller post.

The distance from the tip of the upper pivot down to the lowest part of the hairspring post is measured next (C, Figure 1). This is obtained by placing the hairspring in the balance bridge and leveling it in its natural position so that it is true in the flat. The hairspring may sag in the middle because of the weight of the unsupported collet. The micrometer is then adjusted so that one jaw rests over the upper cap jewel and the other under the collet. The micrometer is turned so that the collet sag is taken up and the hairspring becomes level and flat, simulating the position it assumes when the balance is assembled. The measurement on the micrometer at this point is then noted. Subtracting the thickness of the upper cap jewel (B, Figure 4) from this gauge will result in the height of the hairspring post including the upper pivot (C, Figures 1 and 4).

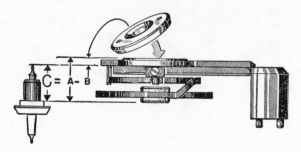

Fig. 4. Securing the height from bottom of the hairspring
shoulder to upper pivot.

To obtain the height of the balance shoulder (D, Figures 1 and 5), sometimes called the riveting shoulder, gauge the thickness of the balance arm or spoke and add to this measurement .10 mm for wrist-watch staffs and .15 mm for pocket sizes. The result will be the measurement (Figures 1 and 5D) of the balance shoulder. The extra .10 or .15 mm are provided to supply enough metal to be riveted on the balance arm.

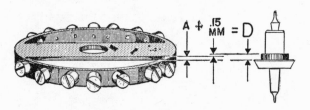

Fig. 5. The proper height of the balance seat.

For the moment, we have all the heights of the staff needed except the balance hub. However, this is easily obtained if we add heights C, D and B and subtract this total from the height over-all (A).

Previously we had determined the distance from the lowest part of the balance to the tip of the upper pivot and the distance from the roller seat at the base of the roller post to the tip of the lower pivot. The connecting distance is occupied by the balance seat. The height of the balance seat also determines the clearance between the balance rim and the pallet bridge.

The heights of the pivots and the thicknesses of the various shoulders or posts and pivots will be found during the actual operations of making the staff.

Where the sample is available, the thicknesses are easily noted by gauging with a micrometer. These measurements should include the pivots, balance shoulder, hairspring post, and the base of the roller post, as well as the tip of the roller post, in order to note the amount of taper needed.

Tools and Materials

The five gravers shown in Figure 6, a micrometer, and the polishing materials and tools illustrated in the chapter, "Adjusting a Balance Staff,"

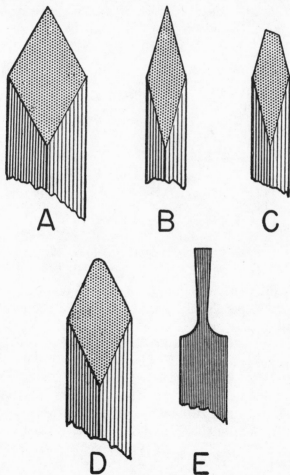

Fig. 6. Types of gravers used in staff making.

are most of the tools needed in this operation. Although the five gravers
pictured are rather helpful, the use of gravers by watchmakers is not uniform. Some watchmakers have a favorite graver to which they have so
accustomed themselves that they virtually use this one alone to perform all
the operations in staff work. Most watchmakers, however, possess a collection of gravers, many of these shaped to suit a particular purpose. The
use and purposes of these gravers will be explained as the need for them
arises.

The material used in making a staff is drill rod or high-carbon steel.
The diameter of the metal should exceed slightly the thickest part of the
staff, namely, the balance hub. This extra thickness will allow for waste
and truing up the metal as it is being "turned." The balance hub should
be slightly smaller in diameter than the width of the balance arm measured
at the balance hole. (See Figure 7.) This must be tempered before being

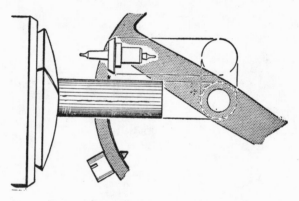

FIG. 7. The thickness of the balance hub.

placed in the lathe. The following tempering process is a standard practice that has had the benefits of long use and found quite satisfactory for
hand-made staffs.

The steel is first heated to a dull red and plunged into cold water or oil.
This tempering makes the metal dead hard. The rod is polished so that the
annealing color may show up better. Polishing may be done with crocus
cloth until some shine appears. The flame that heated the rod is then lowered so that, when the rod is placed over it, the heating process is much
slower. This will help in controlling the amount of heat to be absorbed
by the polished steel rod. The rod may also be placed in the "bluing pan"
that receives the flame. The pan is moved gently over the flame so that the
rod absorbs the heat evenly. At first the rod will turn a light straw color.

Next it will turn darker and assume the color of red wine. Almost immediately after this, the rod will have a rich blue color, similar to that of a blue watch hand. As soon as this color appears, the rod is thrust into the cold water or oil. Steel tempered to this color may be cut on the lathe with sharp gravers, resulting in a fine cut with a polished luster. Staffs cut from this tempered stock need no further tempering.

To permit the metal to be heated beyond the blue color will leave the steel too soft. Beyond the blue color, the steel will lose its luster and turn to a dull gray and, if heated still further, will become red hot again.

Tempering the staff after it is made requires many extra skills and may result in warping the staff. Some watchmakers prefer to temper their staff

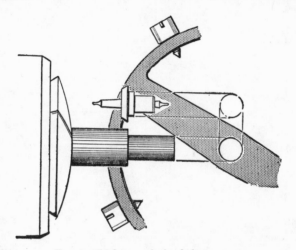

Fig. 8. Thickness of the balance seat.

metal to a wine color and use the carboloy gravers to cut the staff. Although this may be satisfactory for some purposes, it may leave the pivots too brittle and may prevent the staff from being properly riveted as well as damage the riveting punches.

Choosing the proper lathe chuck is important. The correct chuck will give a firm grip on the metal and allow for greater accuracy. Also, it will grasp the metal tightly without undue strain on the draw-in spindle. Most chucks are numbered in tenths of a millimeter. If the micrometer measuring the metal to be used reads 2.40 mm, number 24 chuck should be used. If the metal gauges 1.90 mm, chuck number 19 is proper, and so on.

The stock to be used is placed in the lathe chuck so that it emerges a little longer than the over-all length of the required staff. To permit too much stock to stick out of the chuck will make the cutting operations more difficult because of the possibility of its springing or scratching while the graver is applied to it. This is shown in Figure 7.

The first shoulder to be turned is the balance shoulder (Figures 1 and 8), using graver A, Figure 6. This is cut along the staff from the point where the balance rests, out to the end of the metal. The direction of the cutting always should be toward the lathe. This shoulder should fit the hole in the balance wheel quite snugly without any shake at all. The entire surface must be cylindrical, as illustrated in Figure 8.

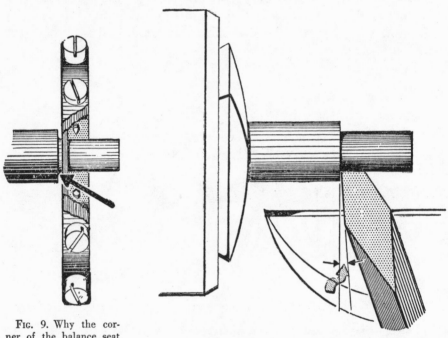

FIG. 9. Why the corner of the balance seat must be square.

FIG. 10a. How a square-corner is cut.

In cutting shoulders, the corners must be sharp or "square." They should not be rounded because the balance arms will not rest firmly against the corners of the shoulder but will lie on the slightly rounded corners as shown by the arrow in Figure 9. In such a case, when the balance shoulder is riveted, the arms will become spread and bent because the slight round will act as a wedge taper. This also applies to the corners at the base of the roller post.

To cut these square shoulders, the graver shown in Figure 6C is used to finish the corners to the desired sharpness. It is a lozenge-shaped tool made over from the same graver as shown in Figure 6B. The top is ground back so that the side and front surfaces present an angle slightly less than 90°. This graver has a strong and durable sharp point.

In using the graver, it is presented as shown in Figures 10A and B. The graver is first used to permit the cylindrical portion of the balance shoulder to reach clear up to the base of the shoulder.

The slight clearance indicated by the arrow is a safety precaution against slight vibrations accompanied with hand turning. To finish, the graver is used to face off the shoulder. The clearance shown by the arrow in Figure

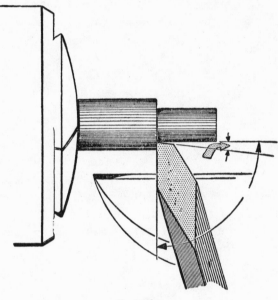

Fig. 10b.

10B prevents the front of the graver from cutting into the cylindrical portion of the shoulder and reducing its thickness. The T rest is held as close to the stock as possible, thus affording greater stability to the graver. Too much pressure should not be applied to the metal as this may result in overcutting or dulling the point. The height of the T rest is adjusted so that the graver is comfortably handled and close inspection with the eye-loupe while the metal is being turned is permitted.

A small piece of white tissue paper clamped at the base of the T rest will give a good visual relief to the metal being cut.

In cutting the staff, light cuts that keep each shoulder cylindrical are desirable and aid in successful staff making.

To obtain the height of the balance shoulder (D, Figure 1), the balance is placed backwards upon the newly cut shoulder so that the arms are snug against the base of the balance shoulder, and the top of the balance faces the lathe as shown in Figure 11. As previously stated, this shoulder should exceed the thickness of the arm by .10 to .15 mm, depending on the size of the balance. As shown by the arrow in Figure 11, the graver then cuts a slight groove to indicate where the balance shoulder ends and the hairspring collet post begins. This shoulder must be undercut so that the extra length may be spread over the edge of the balance hole to secure the staff to the balance. However, this undercut cannot be.made until the shoulder for the hairspring collet (F, Figure 1) has been cut.

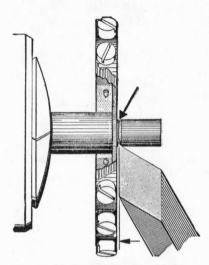

FIG. 11. Marking off height of balance seat.

To obtain the correct thickness of the hairspring post, a set of sewing needles of various thicknesses may be used as gauges. The needles are inserted into the hairspring collet as shown in Figure 12. The hairspring collet shoulder is turned to the diameter of the needle, fitting the collet with the correct grip, and adding about .02 mm for grinding and polishing.

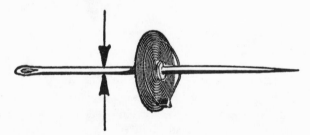

FIG. 12. A set of needles serve as excellent gauges for hair-spring—shoulder thickness.

Cutting this part of the staff is done with the regular turning graver (A, Figure 6) to the determined thickness, leaving the extra thickness required for polishing. Care must be exercised that the cutting of the post does not go beyond the light groove previously made to determine the height of the balance shoulder. As a precaution it would be advisable to cut the hairspring collet post with a square shoulder at its base and later convert it to an undercut.

Undercutting

To undercut the shoulder for riveting, the lozenge graver is used (B, Figure 6), sharpened to a fine point. Although there is very little cutting in making the undercut, this operation requires great care. The graver is applied to the base of the hairspring post. The graver must be held firmly yet applied with little pressure. The front edge of the graver is held almost but not quite parallel to the hairspring collet shoulder. If the graver's front edge were held parallel with the hairspring shoulder or post, the

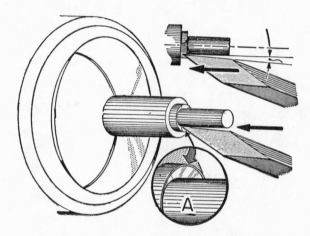

Fig. 13. How the undercut is obtained.

graver might be drawn into it because of the pressure angle of the undercut being made. The pressure of the graver, although light, must be applied in a direction parallel to the axis of the staff as shown in Figure 13.

The undercut should be made to a depth equal to about half the height of the balance shoulder. The angle of the countersink is determined mainly by the difference in thickness of the balance shoulder and the hairspring post. The angle is cut from this depth at the hairspring shoulder outward to the edge of the balance shoulder.

To insure against shortening the height of the balance shoulder, the countersink should not reach up to the very edge of the balance shoulder but stop just short of it, leaving a slight flat portion visible as shown by the arrow in A, Figure 13.

Cutting the Pivots

The round-tipped graver shown in D, Figure 1, is used to cut the pivots. Balance staff pivots are made up of two distinct sections: one a true cylinder, the other conical, cut and shaped to blend perfectly into the cylindrical portion of the pivot so that it looks like one continuous piece. The proportions of the pivot are shown illustrated in Figure 14. The cylindrical part of the pivot is generally one-third as long as the conical section.

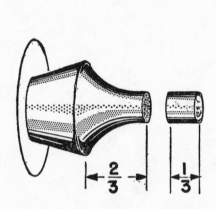

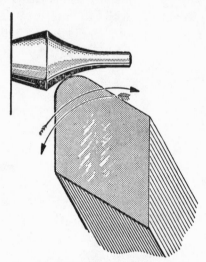

Fig. 14. Proportions of cylindrical pivot-end to conical section.

Fig. 15. Cutting the conical section.

In cutting the pivot of the staff, the round-tipped graver is used in a series of slow short strokes, curved to the desired shape of the pivot as shown in Figure 15. It is advisable to make the pivots just a little longer than needed since these can be shortened to suit later adjustments. Of course, the pivot is left oversize for the subsequent polishing operation.

Oil Grooves

The purpose of the oil groove below the coned section of a pivot is to trap the excessive oil that otherwise might run down the staff to the hairspring. The nature and angle of the undercut keep the oil in its corners because of capillary action. This is shown in Figure 16. This groove is cut with the undercutting graver. First, a grooving cut is made and, then, a facing cut. The height of this groove is generally about one half the length of the entire pivot. However, this groove may be made longer for the sake of appearance but it should not be made so long that the hairspring post becomes shorter than the height of the hairspring collet. The groove must not be cut too deeply, otherwise the staff might become weakened at this point.

The edge of the hairspring shoulder is beveled slightly to permit the hairspring collet to enter upon it.

At this point in making the balance staff, it is advisable to polish the pivot, hairspring post, and oil groove because the subsequent operations will weaken this part of the staff and thus hinder the polishing operations. Instructions on polishing the pivots and shoulders of the staff can be found in the chapter dealing with "Adjusting the Balance Staff."

In polishing the hairspring shoulder, care must be taken not to reduce the

Fig. 16. The oil groove traps excess oil.

height of the balance shoulder. A staff that has been cut skillfully with cylindrical surfaces just about .02 mm above the finished thickness should not require much polishing to bring the staff to a fine black polish.

After polishing the pivot, oil groove, and hairspring post, the balance hub is cut. This should be done so that it slopes back in an angle of about

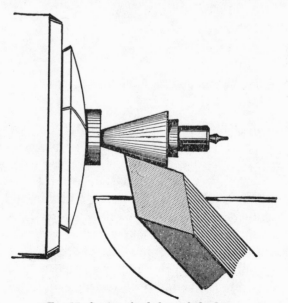

Fig. 17. Cutting the balance hub slope.

30° and cut almost the full length of the metal protruding from the chuck
as shown in Figure No. 17. This slope is cut until just a little of the
cylindrical or original thickness of the metal is left. It also should be mentioned that this cut must be made flat without any irregularities in its surface because, immediately after this operation, the slope is polished.

The roller post is cut while the metal is still in the lathe. It should be
pointed out that the metal must not be removed from the lathe during the
whole operation of making the staff, otherwise the truth of the various surfaces may be spoiled. A staff cut while in one chuck without being re-

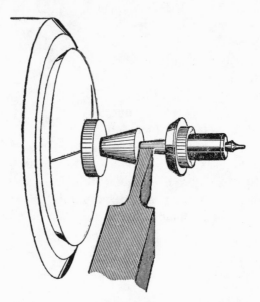

FIG. 18. Cut-off tool is handy in turning the roller
post.

moved will remain true since it is turned on only one axis. The graver to
be used in cutting the roller post is the one shaped like a parting tool
(Figure No. 1 "E"). This graver is used because its shape permits it
to cut the roller post in the cramped space between the balance seat and
the lathe chuck. Figure 18 shows the tool performing this operation.

Measuring the diameters of the exposed section of the staff is comparatively simple since these measurements may be conveniently obtained by
the micrometer. Because of the limited space of the roller post, this thickness is measured with calipers. However, a micrometer with thin jaws may
also be used. A millimeter caliper with vernier markings is shown in Figure 19. This has readings calibrated in tenths of a millimeter. The vernier

markings on the indicator permit readings in hundredths and have a capacity of 12 millimeters.

The roller post should be turned to a slight taper. Generally in a staff, such as used to illustrate this chapter, the taper is about .03 mm thicker at its base than it is at the beginning of the taper. Using very light cuts will result in a smooth, fine finish that requires little polishing.

After the taper is turned to a diameter within the limits previously discussed to allow for polishing, the cone and pivot are cut.

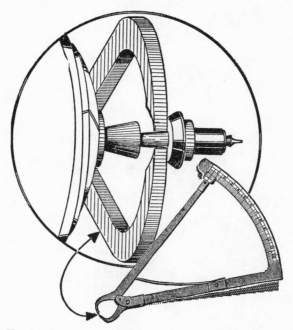

Fig. 19. Type of gauge used to measure unfinished roller post.

Cutting the cone and pivot at this end of the staff is done with the round-tipped graver used previously to cut the pivot at the hairspring end of the staff.

It should be remembered that up to now the staff was connected to the metal in the lathe chuck by the sturdy roller post. However, when the pivot is cut, it thins out this link. If any strain were placed upon the staff, it might then break off at this point. Therefore, in forming this back pivot, the graver should cut from the roller post toward the chuck. In this way the metal that is being made thinner becomes part of the extending staff while the graver working inward places a strain only on the thicker metal extending from the chuck.

The cylindrical part should be made as long as possible before parting the metal. This is shown in Figure 20. The staff may be cut off from the metal in the chuck by the gravers shown in Figure 1, A or B.

When the staff is parted, it is placed in a chuck that fits the hairspring post as shown in Figure 21. Selecting the chuck should be done by following the directions previously given on that subject. It is important that the

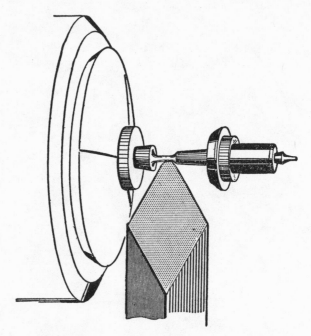

Fig. 20. Roughing out the back (lower) pivot.

proper chuck be used if a snug fit and accuracy are to be expected. The staff may then be finished, giving such finishing touches to the base of the roller post and polishing the roller post and the pivot, as shown in Figure 21. Cutting as much of the staff as possible while it was in the first chuck relieves a great deal of strain and possible inaccuracies while the staff is later held by the hairspring shoulder.

If when fitting the hairspring shoulder in the chuck it is impossible to make it run true, there is indication that the hairspring post was not turned cylindrically but rather to a taper. Or else the chuck selected is at fault. An oversize chuck should not be used because it will grip the metal only at the point where it enters the chuck. Furthermore, the chuck is made to contract too much with the result that it may become wedged in the lathe.

Forcing the metal into a smaller chuck spreads the chuck and the metal is gripped only near its innermost part. In both cases the metal will not run true and the chucks may become sprung and unfit for further use.

If no accurate chuck is available to finish the staff, the job can be done very effectively by placing it into the hollowed-out coned section of a brass

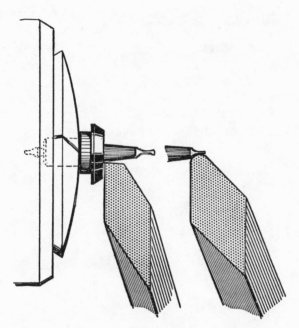

Fig. 21. Finishing the staff in a split-chuck.

rod. The staff is held in place by melted shellac. This method is called the "shellac chuck" method.

To prepare this "chuck" for the staff, a piece of brass rod about 4.00 mm is placed in the lathe chuck that will accommodate it most accurately. Most lathes are equipped with "cement brasses." These are threaded rods of brass about 5.00 mm thick and threaded at one end. The threaded ends fit into a special chuck, threaded also to receive these brasses. Others are tapered instead of threaded and have a small pin projecting from the side which fits into a slit into the taper-holed chuck. These are shown in Figure 22.

Whatever the style of chuck used, the brass part is prepared for staff making in the same way. The lathe graver is applied to the front center of the brass rod and a deep conical countersink is cut as shown in Figure 23 in a sectional view. This countersink must come to an absolute point if

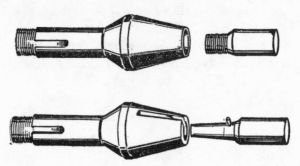

FIG. 22. Types of cement chucks.

the final operation is to prove successful. If the countersink should appear as in Figure 24, this inner point must be removed by cutting with a forward pressing movement of the pointed graver.

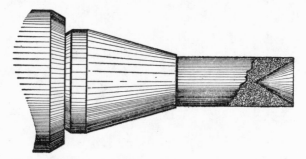

FIG. 23. Cut-away view of correctly turned cement-brass.
Countersink must come to absolute center.

A simple test to prove that the countersink has been cut to an absolute centered point without the bead shown in Figure 24 is to turn the lathe

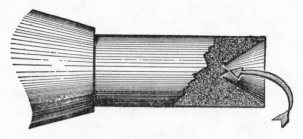

FIG. 24. Point indicated by arrow must be turned away so
that it appears like figure 23.

while a very sharp-pointed steel pin or needle is made to rest in this countersink. If the pin seems to wobble, it is a positive indication that it is resting beside a small bead raised by the graver while it was cutting off center. If the needle has no perceptible movement, then it may be assumed that the countersink has a dead center and is ready for the next step. The rod now has a center cut regardless of its position in the lathe chuck and this must *not* be removed from the chuck for the remainder of the operations.

The brass rod is heated by an alcohol lamp flame until the stick shellac that is placed in the countersink melts and completely fills the conical cavity. The lathe is turned slowly so that the heat is evenly distributed. While the shellac is still soft, the staff is placed into the countersink, pushing it carefully through the semi-liquid shellac.

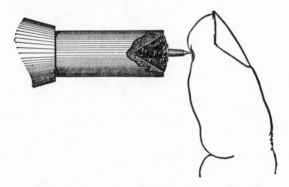

Fig. 25. Staff gently pressed into soft shellac with finger tip.

The tip of the index finger is pressed against the end of the staff extending out of the small mass of shellac as shown in Figure 25. This will push the staff into the countersink until the pivot rests against the pit of the countersink. The shellac must be soft enough to permit this adjustment to the staff. Should the shellac become hard before this adjustment is made, heat is reapplied. Care must be exercised not to apply too much pressure against the pivot or else the opposite pivot inside the shellac may become broken because of the extreme pressure against the brass countersink. For this reason, the fingertip is recommended, being more sensitive to the pressure required.

When the staff seems to indicate that the pivot is resting at the base of the countersink, the pressure of the finger is removed and a piece of pegwood is held lightly against the roller post while the lathe revolves slowly.

This will true up the staff. This application of pegwood against the roller post must be continued with delicate pressure until the shellac has become hardened (Figure 26). Observation of the protruding staff and pivot will show that the staff is held securely and very accurately centered.

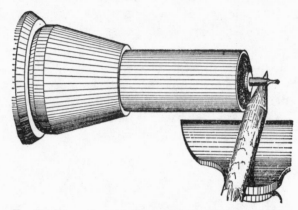

Fig. 26. Pegwood centers the staff in the hardening shellac while lathe is slowly turned.

After this, the staff may be finished, care being taken to take only very light cuts if any part of the extending staff still needs shaping. The roller post and the roller seat as well as the balance pivots may then be polished and finished.

The brass chuck may now be removed from the lathe and is reheated at a point farthest from the staff to soften the shellac. This precaution is necessary to protect the staff from the heat and guard against softening the finished product. When the shellac becomes soft, the staff is pulled out and dropped into a phial of alcohol where the shellac will be dissolved.

If upon gauging the staff it is found that further adjustments may be necessary, these may be made, using the methods described in the chapter dealing with adjustment of the balance staff. If rechucking is necessary, the shellac chuck method may be used again. Testing the staff prior to fitting it to the balance may be done using the methods described in the chapter, "How to Match a Balance Staff."

SUMMARY

1. When a staff is unavailable or the nearest match requires excessive alteration, it is better to make the new staff from the raw stock.

2. When the sample is present, this is gauged for measurements and recorded on a sketch of the staff.

3. If the sample is lost, the measurements can be obtained from the movement, balance roller, and hairspring.

4. The metal used to make the staff is high-carbon steel annealed after hardening to a blue color.

5. The height of the balance shoulder is generally .10 to .15 mm higher than the thickness of the balance arm.

6. The thickness of the hairspring collet post may be had from the thickness of a needle that fits into the collet with the correct tension.

7. The oil grooves are made to trap excessive oil that runs down from the pivot.

8. The upper pivot, hairspring post, and hub are polished before the roller post is cut.

9. Shellac chucks are used to hold parts to be turned that cannot be held accurately and securely in regular split chucks.

10. Only light cuts are suggested when finishing the staff while it is extended from the shellac chuck.

QUESTIONS

1. When is it necessary to make a staff?
2. How is this metal tempered in preparation for the staff making?
3. What is the method of choosing a chuck?
4. Describe the five gravers recommended for the operations in staff making.
5. How is the "height over-all" obtained if the sample is missing?
6. How is the height of the roller post noted without a sample?
7. How is the height of the hairspring shoulder and pivot obtained?
8. How is it possible to obtain the thickness of the hairspring post without a sample?
9. Why should the shoulders and pivots be left oversize after cutting with the graver?
10. When the staff is turned around in the lathe and the staff will not run true, what are some of the reasons for this?

IX

ADJUSTING A BALANCE STAFF

PART I: How to Make the Balance Staff Shorter

When selecting or matching a balance staff, it is sometimes impossible to duplicate it. In that case, a staff should be chosen whose measurements most nearly match the original. However, the parts that do not match must be selected larger than those of the sample so that they can be reduced to exact size.

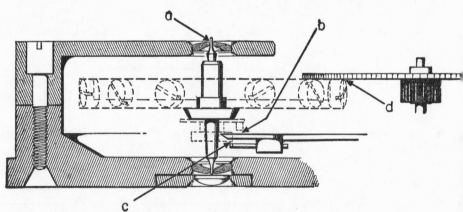

Fig. 1. Observing the staff prior to making adjustments to it.

When such a staff is chosen, it should be placed into the movement as though it were attached to the balance wheel. The balance bridge cap jewel (endstone) should be removed as a safety precaution, permitting the pivot,

150

if it is higher than needed, to protrude through the hole jewel. This also prevents breakage to the pivots or cap jewels. In this way the staff may be easily observed for alignment.

Also, despite many efforts to obtain a satisfactory duplicate, sometimes a staff is fitted that needs adjustments. These adjustments should be made only after very careful observation.

In Figure 1, the staff is shown placed into the movement; the balance bridge, minus its cap jewel, is in place. This staff is too high as indicated by the pivot sticking up through the balance bridge hole jewel at (a).

Before making the staff shorter, an examination is made to determine where the correction should be made and which pivot, upper, lower or both, has to be shortened. In this case, the balance and roller are illustrated in phantom view. The roller is shown, staked to the staff, but its position on

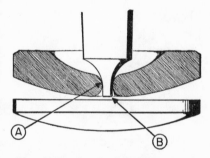

Fig. 2. Result of shortening a pivot without cutting into the cone.

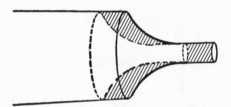

Fig. 3. Shaded area indicates metal to be removed when shortening a pivot.

the roller post and roller seat of the staff indicates that it is too high for the roller jewel to be engaged by the pallet fork (b). The guard finger is also out of line with the safety roller (c).

The balance wheel, when staked to the staff, will move in a plane too high and will scrape the center wheel should they overlap, (d, Figure 1).

In order to give clearance to the balance and bring the rollers into engagement with the pallet fork and guard finger, the lower pivot must be shortened. This will lower the entire unit and permit the upper pivot to descend in the upper hole jewel to its proper level.

However, if the pivot is simply shortened, it cannot go through the hole jewel for sufficient distance to rest upon the cap jewel. In Figure 2, the lower pivot was shortened to comply with the demands of Figure 1. However, now the coned section of the staff is resting against the edge of the hole jewel at A, Figure 2. The pivot end does not contact the cap jewel

(B, Figure 2). This will not permit the watch to run because of the wedge-like shape of the pivot and the resulting friction.

To permit this pivot to reach the cap jewel, a portion of the coned section of the staff will have to be cut away so that the cylindrical portion of the staff will be long enough to reach the cap jewel. In Figure 3, the shaded area represents the part of the staff that must be cut away to make the remaining portion resemble the unshaded area. In fact, the shortening of the cone should be done before reducing the pivot length. In this way, the original length of pivot serves as a guide for the thickness of the final

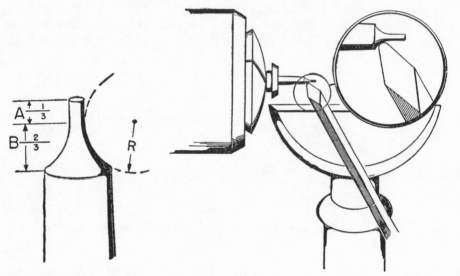

FIG. 4. Acceptable pivot pro-
portions.   FIG. 5. The rounded tip graver is used to cut conical points.

desired pivot as well as aiding to keep the new length cylindrical. The extra-long pivot which results from cutting the cone into the body of the staff is reduced later in the final operations.

A well-shaped pivot should resemble the one pictured in Figure 4. Here the pivot (including the cone) is considered as one unit. The cylindrical portion A makes up one-third the length, and the coned section B the remaining two-thirds. The radius as shown by the arrow R indicates how the coned section should be shaped. A pivot thus formed will be strong and serviceable.

To cut the cone shorter and yet retain the radius or curved effect, a graver whose point has been slightly rounded off is used. Such a tool is shown in Figure 5, cutting the coned part as specified in Figure 3. The graver must be well sharpened with a smooth, even edge so that the resultant surface will also be smooth, requiring little to finish and polish the cone or pivot. The lathe chuck that will hold the staff must be true and close fitting, otherwise the new pivot may not be true and on the same axis as the pivot at the other end of the staff. This would throw the balance out of round and out of poise, making good timekeeping almost an impossibility. Therefore, before any cutting operation on the staff is started, the staff must spin perfectly true in the lathe.

The original (cylindrical) portion of the pivot, as previously stated, remains on the staff to the last as a guide to the thickness of the pivot. This also serves as an aid in guiding the grinding operation to preserve the cylindrical shape of the pivot.

When the coned section has been cut into the body of the staff, the new pivot should be left slightly thicker than needed so that this may subsequently be reduced to size by successive grinding and polishing operations. The staff is now ready to be reduced to size.

Part II: How to Reduce the Thickness of a Balance Staff Pivot

Balance staff pivots may be reduced in thickness by grinding and subsequent polishing. A pivot may have to be reduced in thickness when it binds in the jewel hole or has not sufficient freedom (sideshake) in the jewel hole. How to test for correct pivot thickness has been explained previously in the chapter "How to Match a Balance Staff," illustrated in Figure 30 of that chapter.

Although there are many methods used to reduce pivots to size, the most popular utilizes a triangular Arkansas oilstone slip applied to the pivot while the staff, or balance and staff, is held in the lathe. Other methods are: the use of special pivot lathes or turns; use of polishing shovels; and the use of the transverse grinder, sometimes called the "pivot-polisher."

To use the Arkansas oilstone slip, the staff is placed in the lathe and the oilstone slip is applied to the bottom of the cone and pivot of the staff. A slight smear of watch oil is put on the stone to aid in this operation. The lathe is turned at a slow speed towards the operator while the oilstone slip is worked back and forth. This is shown in Figure 6. Care must be taken that the pivot is ground into a cylindrical shape. The lathe is turned slowly so that the reduction of the pivot thickness may be slow, gradual, and controlled. This will permit frequent observation and possible correction in the manner of holding and applying the oilstone slip.

Pressure against the pivot must be light during this operation so that better control may be had over the shape of the pivot. During this time, the eye loupe should be focused at the top of the pivot length so that any discrepancies may be quickly noticed and the method of holding the slip may be altered to suit the shape desired. The oilstone must also be held so that its flat top is parallel to the axis of the pivot as shown in Figure 7.

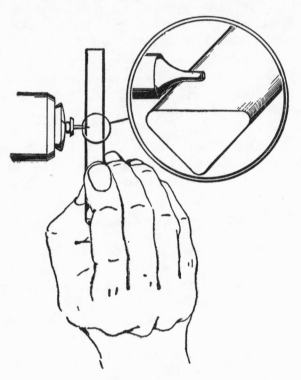

Fig. 6. How an oilstone slip is applied to the pivot to make it thinner.

If the oilstone slip is pressed too heavily against the pivot, the oilstone may exert an uneven pressure upon the pivot so that the pivot will be pointed rather than cylindrical. Such a fault is illustrated in Figure 8. On the other hand, if the oilstone slip is applied with too much emphasis upon the conical part of the pivot, the result will be a bull-headed pivot, as shown in Figure 9. As previously stated, the eye loupe is focused upon the top of the staff and the fingers controlling the oilstone slip co-ordinated with the eye to manipulate the slip in producing a correctly shaped pivot. Some practice with old, broken staffs, both in cutting the cone with the graver and lengthening the pivot as well as the use of the oilstone slip, will more than repay for the time spent in such efforts.

Where the cone taper or radius of the staff (Figure 4, R) must be longer, the oilstone slip is still held with its flat top parallel to the axis of the pivot,

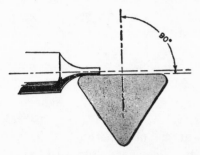

FIG. 7. Hold top of slip parallel to axis of staff for cylindrical pivot.

FIG. 8. Improper application of oilstone slip results in pointed pivot.

but the oilstone slip may be twisted to one side as in Figure 10, so that the axis of the staff and the axis of the oilstone form an angle greater than 90° (right angle) to the amount corresponding to the cone shape desired.

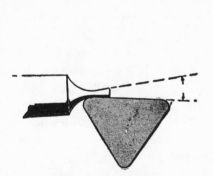

FIG. 9. Another reason for improperly shaped pivots.

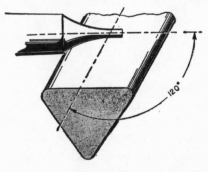

FIG. 10. Twisting the oilstone slip sideways while keeping the top surface horizontal will tend to lengthen the conical section.

The pivot should be reduced until it is almost down to size or about .01 mm from the final thickness, cleaning the pivot often with pithwood to observe clearly the results. Because the finish on the pivot left by the application of the oilstone is far from smooth, this extra thickness is removed in successive applications which will further reduce the pivot and polish its surface at the same time. For this next step, a jasper slip is used.

Jasper stones are generally hard, red stones impregnated with ferrous oxide, similar to a material called jewelers' rouge. This stone, used dry, is manipulated in the same manner as the oilstone. It will reduce the pivot

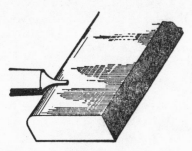

more slowly and give the pivot a slight polish as well. The surface of the jasper stone selected should be flat and smooth, not coarse. Sometimes, when using the jasper slip, the surface may become clogged with the residue of the steel pivot. This residue may be removed by cleaning it with a piece of chamois or rubbing the stone across the knuckle of the thumb. The jasper stone is manipulated until the pivot is almost down to

Fig. 11. Burnishing a conical pivot.

size. The final finishing is done with the steel burnisher. Figure 11 shows this tool in operation.

Part III: How to Polish a Balance Staff Pivot

Using a Pivot Burnisher

The steel burnisher has two differently shaped edges. One of these edges is rounded off for finishing pivots with conical roots. The other is sharp-edged for finishing pivots with sharp, square shoulders. The burnisher is used with a smear of light watch oil and applied in the same manner as the stone slips. The results will be a highly polished, strong pivot.

The burnisher can be made to reduce pivot thickness by finishing its surface with emery paper, making the grain run across the burnisher from edge to edge as shown in Figure 11. While this reduces the thickness of the pivot, it polishes it at the same time. When refinishing the burnisher, the emery paper is glued or tacked down to a firm, flat surface and the burnisher is rubbed across the emery in straight, steady strokes.

The burnisher may also be grained by doing the same thing, using a carborundum stone, however, as the base. The minute, fine grain produced in the tool acts as a superfine file.

Polishing a Dirty or Tarnished Pivot

Sometimes a pivot is encountered whose surface is either dirty or tarnished. Burnishing may reduce the thickness of the pivot, and therefore the use of any kind of active abrasive would mar the fit of this pivot in the hole jewel. A very high luster and polish may be given this pivot without reducing its thickness perceptibly by using a piece of pegwood charged

with a mixture of oil and jewelers'
rouge. This is shown in Figure 12. The
pegwood is prepared by cutting two
flat sides to form a slight, acute angle.
A little watch oil is rubbed across the
flat top of the pegwood. Then a stick of
jewelers' rouge is scraped lightly to
present a clean surface free from foreign matter. This cleansed surface is
rubbed over the oiled pegwood so that
the pegwood becomes charged with the
rouge.

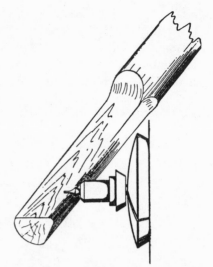

The rouged pegwood is then run
across the pivot. Because the pegwood
is much softer than the pivot, it will
soon shape itself into the pivot, cover-
ing the whole surface and polishing it
at the same time. The pivot is then

Fig. 12. A piece of pegwood charged
with jewelers' rouge and oil brings up
a high lustre on the pivot.

cleaned by running it into a piece of pithwood. The pivot is then tested for
thickness, length, shape and fit.

Part IV: The Use of Pivot Lathes

For quick efficient "polishing" of pivots to size, the pivot lathe is un-
equaled. The small pivot lathe, shown in actual size in Figure 13, is eco-
nomical, simply constructed, and easy to operate. Because the pivots of
the balance staff rest on their own centers and cones, accuracy is reasonably
assured. Errors introduced by imperfect lathe chucks and wobbling pivots
on such chucks are thus avoided.

This tool consists of a frame with two legs which fit into a metal base
A, Figure 13. This base may be screwed permanently to the edge of the
bench so that, when the tool is not needed, the frame is lifted out of the
base and put away. The frame B has a small headstock into which is fitted
the rod C. This rod has a button at the outer end. The other end of the
rod has a hole drilled and cross-holed for the reception of a balance pivot.
The rod may be advanced and locked in any position by the top set screw
D. Upon the rod C the pulley wheel (B, Figure 14) is placed loosely. At
the other end of the tool is the disc holder E. This holder is fastened by
the nut F.

To use this tool, the "fiddle bow" (A, Figure 14) is twisted over the pulley wheel. A disc is chosen whose hole will permit all the cylindrical portion of the pivot to be "polished" and a slight part of its cone to stick through. Some tools have three holes in one disc. In this case the suitable hole must be "aligned" on the same axis as the rod C. The balance pivot E is placed into the hole in the disc F. The rod is then advanced so that

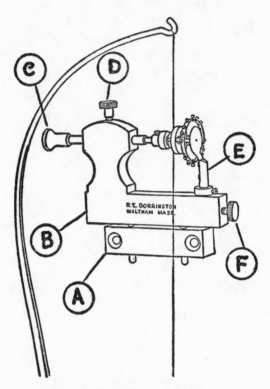

FIG. 13. Pivot lathe (actual size).

the pivot D will enter through the hole. This rod is brought up close so that there is no endshake of the balance between the rod hole and the disc hole. The nut D, Figure 13, is then fastened, securing the rod in place. The hairpin-shaped wire, placed through the pulley wheel B, is pushed forward so that its pins, G and H, straddle a spoke of the balance wheel. These pins should not be advanced so far that they touch or scrape disc F. By pushing and pulling the bow A back and forth, the pulley wheel is made to turn upon the rod C. Since the straddle pins G and H are attached to this pulley wheel, any action of the fiddle bow will transmit motion directly to the influenced balance wheel, and thus the pivot sticking out of the hole F is ready to receive the burnisher or jasper stone. An Arkansas oil-stone should not be used; it sheds abrasive powders which find their way into the disc hole, altering the cone and enlarging the hole.

To reduce the thickness of a balance pivot with this tool, a jasper stone or a pivot burnisher, J, is held up against the underside of the pivot. The bow and the jasper slip or burnisher are pushed and pulled in the *same* direction. This will produce a counter action of the pivot across the bur-

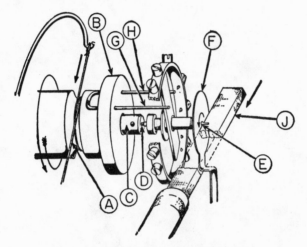

Fig. 14. Close-up detail of the pivot lathe.

nisher, resulting in a more rapid reduction or polishing of the pivot with satisfactory results.

Care should be exercised that the pivot file or burnisher be held against the pivot with the same precautions indicated in Figure 7, avoiding the errors illustrated in Figures 8 and 9.

As previously stated, oil is used with the burnisher but not with the jasper slip.

Although this tool is fairly simple to operate, some practice should be had employing old balances and pivots. Using this tool but a little should provide a proficiency equaling that obtained with the same tool in actual factory work.

The "Jacot" Pivot Lathe

The tool shown in Figure 15 is known to English-speaking watchmakers as the "Jacot" tool. This tool undoubtedly offers one of the very best methods of reducing or polishing pivots to size and shape. With a correct knowledge of its use and some practice, correctly shaped and dimensioned pivots are practically guaranteed.

This tool, shown about three-fourths its actual size in Figure 15, is held in the vise. A fiddle bow, the same as used in Figure 14, made of spring

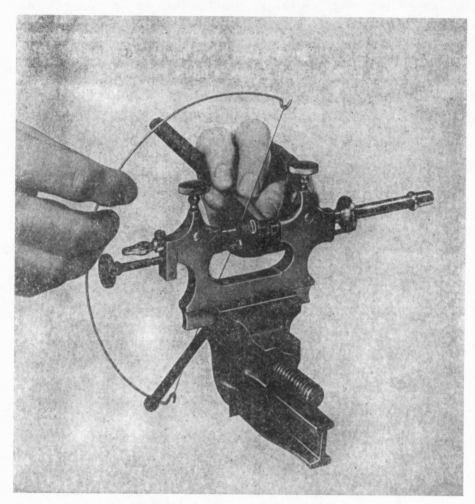

Fig. 15a. The Jacot Pivot lathe in use.

wire about 2.00 mm thick and horsehair, is wound once around the pulley A. The pivot to be straightened, reduced, or polished is placed in a groove marked with the gauge of the pivot size desired. The hardened, grooved spindle E, Figure 15, is brought to the uppermost position and locked in place. The opposite balance pivot is placed into the hole in the pulley rod and this is advanced until *only* the cylindrical portion of the pivot rests in the groove.

Choosing the proper groove is of importance if a well-proportioned cylindrical pivot is desired. Although the gauges marked adjacent to the grooves serve fairly well, the pivot should be placed into a groove that permits about one-eighth the pivot diameter to stick up above the flat surface of the spindle as shown in Figure 16. This will permit the pivot A to obtain a firm seating in the groove C and allow the burnisher B access to the pivot's outer circumference as well as to permit it to obtain guidance from the adjacent flat-surfaced guide D. For instance, if a pivot is .12 mm

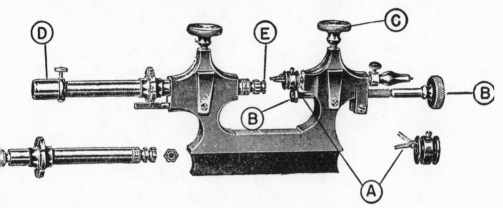

Fig. 15b. The Jacot tool (pivot lathe) in detail.

thick and it is desired to reduce this to a .10 mm thickness, a groove No. 10 might be chosen. However, if a pivot, let us say, is .14 mm thick and it is desired to bring this down to .10 mm, then it should first be placed in the No. 12 groove and, after assuming that thickness, it is again transferred to the No. 10 groove for the final reduction and polish. To attempt to fit too thick a pivot in a small groove may invite difficulties in keeping the pivot seated in the groove and make it equally difficult to obtain a well-shaped pivot.

The principle of this tool is that it allows one pivot to rest in the pulley rod while the pivot to be adjusted rests in the groove which is precisely on a line of centers with the tool. The groove is ground to a depth indicated by the gauges marked adjacent to them. The surface into which these grooves are indented are perfectly flat with a corresponding flat surface extending beyond the groove. Thus a burnisher or pivot file may cut a pivot until its diameter is reduced so that it is exactly on level with the hard, flat surface. The burnisher cannot reduce it any further, insuring a pivot of exact dimensions. Also since the groove is cut straight along with the lines of centers and also since the top edges of the grooves are perfectly flat, the burnisher cannot help but cut a pivot that is also cylindrical. The burnisher used must have a rounded edge as illustrated in Figure 16 in order to fit into the shape of the conical root of the pivot, polishing it as well.

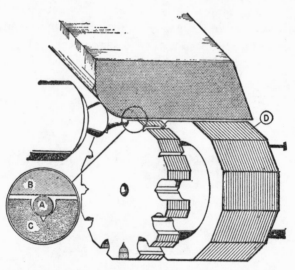

Fig. 16. The portion of the pivot's diameter (*A*) above the surface of the groove (*C*) will be reduced by the burnisher (*B*). *D* serves as a guide for the burnisher.

The Burnisher

Burnishers come with two ends as shown in Figure 17, ¾ actual size. One end has fine filing surfaces and edges. This end may be used to reduce the pivot thickness to within .005 mm of the desired thickness. The filing end has two opposite edges, one with a sharp edge for filing pivots with square shoulders such as used with third wheel pivots shown at right in Figure 18. The diagonally opposite end has a rounded filing edge which is used to work upon the conical portion of balance style pivots.

These burnishers come with the sharp angles in two styles—right and left. For right-handed watchmakers who rest the burnisher on top of the pivot, a *left*-handed burnisher is used. Where the work is held under the pivot, a *right*-handed burnisher is used. It is advisable to have both types.

When the burnisher is used to finish or reduce pivots, it is smeared lightly with a coating of fine oil. No oil is used when the filing surfaces of this tool are employed. The oil keeps the fine steel particles, shed from the pivots, floating clear of the crevices in the burnisher. This keeps the surface clear and clean.

FIG. 17. Combination pivot file and burnisher.

To reduce the length of a pivot, the holed-disc-ends are used. The pivot is placed through a hole in the disc and the pivot file is used as shown in

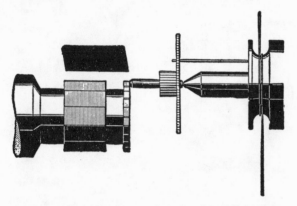

FIG. 18a. A burnisher purposely inclined for left hand use.

Figure 19. Here the file or burnisher is rested upon the cylindrical recess of the spindle and the pivot file is pushed back and forth against the pivot-

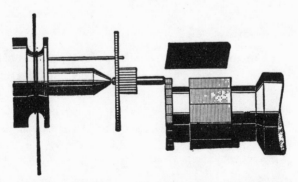

FIG. 18b. A burnisher with a rake designed to be used with the right hand.

end until the adjustment desired is obtained. However, this operation generally places a burr on the pivot end. This must be removed with the burnisher run across the edge of the pivot while it is rotated for a few turns as shown in Figure 20. Because these holed discs are very thin and brittle,

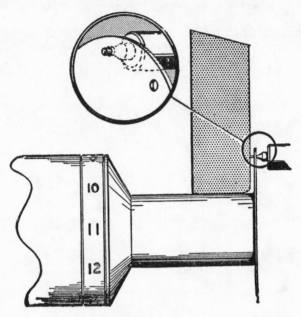

FIG. 19. Burnishing pivot ends.

care must be used that no strain be placed upon them as they might chip off. When not in use, they are protected by the scabbard D, Figure 15. The hole chosen in the disc for this purpose must permit only the cylindrical portion of the pivot to stick through it.

Polishing Shovels

In the reduction of round surfaces such as pivots, roller posts, hairspring shoulders and balance shoulders, conical surfaces, or oil grooves, the polishing shovels used with the lathe are efficient and render a most satisfactory polish. The shovel shown in Figure 21A is used in conjunction with an eccentric tailstock taper B, also shown in this illustration. The shovel illustrated is made of iron, hard brass, or copper pipe. A section of such pipe about $1\frac{1}{2}''$ long and about $1''$ diameter is cut along its length resulting in two U-shaped sections. These are soldered or screwed to thin brass rods about $\frac{1}{8}''$ diameter and about $4''$ long, to the underside of each section. The top surfaces are then filed along their lengths so that these surfaces are straight and parallel to one another. When this is done, the surfaces are "grained." This is accomplished by "cradling" the rounded bottom of each shovel in the left hand between the index and middle fingers

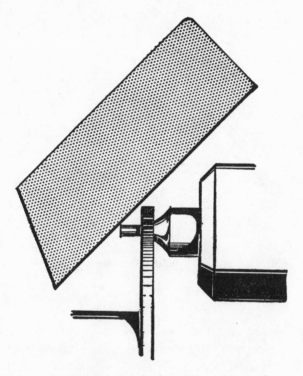

FIG. 20. Removing the burr from a pivot-end.

as shown in Figure 22. In the right hand, a fine file, preferably a No. 6, is placed across the top of the shovel at right angles to its length and straight, steady strokes are used in one direction across the tops. This produces a fine grain.

Fine particles of polishing or abrasive powders, when prepared and placed onto the shovel's surface, will partially imbed themselves into the microscopic grooves created by the file and thus the surface becomes an abrasive or polishing lap. The burrs created at the edge of the shovel should not be removed, finding useful purpose in polishing the innermost corners of square-shouldered or even conical pivots.

The shovel lap is kept steady and upon a given and adjusted keel by the fact that one surface is steadied by the object in the lathe chuck, and the opposite surface B, Figure 23, rests up against the rollers set into the tailstock taper. The tailstock taper B, Figure 21, is constructed by turning a brass or steel taper to fit the tailstock accurately. Then a small disc is

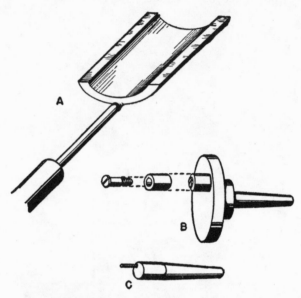

FIG. 21. Details of a polishing "shovel."

mounted permanently onto the taper. Somewhere near the edge of this disc, a hardened steel pin is secured. In more elaborate set-ups, a steel roller is mounted to ride upon a screw. A very simple tailstock arrangement is made by taking a regular tailstock taper, filing the end flat and

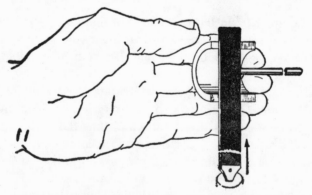

FIG. 22. How the shovel is held while its surface is being grained by the flat strokes of a file.

drilling a small hole near its edge into which is driven a hardened steel pin as shown in Figure 21C.

If the diameter of a cylindrical surface is to be reduced or polished, the shovel's surface A, Figure 23, is placed under and against the shoulder to be worked, and the opposite side of the shovel B is placed under the tailstock roller. The tailstock spindle is turned until the surfaces of the shovel,

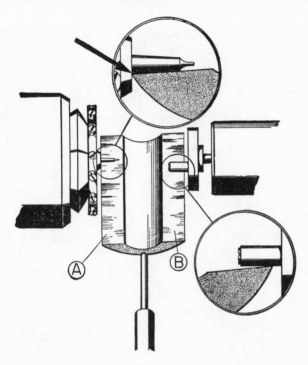

FIG. 23. How the shovel is applied and adjusted.

held in the left hand, are on a line parallel to the axis of the lathe. A slight sideshake is permitted between the tailstock disc and the shoulder of the object held in the lathe. When this is accomplished, the tailstock spindle is locked in place and the flat surface of the shovel is ready to receive the powders.

The abrasive powders used for the reduction of pivots or round surfaces may be fine Arkansas oilstone powder or fine lavigated aluminum oxide. To prepare these powders for use, a drop of fine oil is placed upon a clean glass or steel surface. Upon this oiled spot is placed a scoop of these powders about 1/4″ in diameter. The powders and oil are mixed thoroughly with a steel spatula as shown in Figure 24B. The spatula is turned over to crush any of the mixture that might have clung to its edges, occasionally gathering material that has spread away from the center of the mixture and crushing it again. Oil or powder may be added until the mixture becomes thoroughly and evenly mashed, with the consistency of putty.

A simple spatula can be made by taking a piece of brass wire about $\frac{1}{8}''$ thick and 6″ long. One end of this rod is slit with a fine jeweler's saw

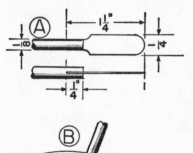

for about $\frac{1}{4}''$. A piece of mainspring about $\frac{1}{4}''$ wide by $1\frac{1}{4}''$ long and quite thin is inserted into this slot and soldered into place. The end tip of the mainspring is rounded as shown in Figure 24A. Both ends of the rod may be thus fitted; one for use with the abrasive powders, and one for mixing the polishing compounds.

To grind a balance pivot with the shovel, it is aligned with the tailstock and pivot so that both surfaces are level. A slight smear of abrasive putty about the size of the head of a bridge screw is spread by the spatula along the length of the flat edge that is to rest under the pivot. With the lathe turning toward the

FIG. 24. Details for construction of a spatula and its application.

operator, the shovel is pushed back and forth with slow, even strokes. It must be emphasized that pivots become reduced in thickness quickly by this method. Therefore they should be checked often for size.

To polish a pivot, the same procedure is followed as with grinding. However, a mixture of diamantine is made, using a clean and different glass or steel plate for the mixing and another spatula. This precaution is necessary so that no grit or abrasive is introduced to the polishing compounds. The staff pivot is cleaned by having it run into a piece of clean pith until examination shows it to be clean.

If possible, another shovel should be used for polishing. If not, it must be thoroughly cleaned with benzine and grained again with the No. 6 file. The polishing mixture is placed upon the flat surface of the shovel. Working this under the pivot for a few strokes will give it a fine luster, insuring a straight, cylindrical pivot.

Balance shoulders, hairspring shoulders, and roller posts may also be reduced to size and polished in the same manner.

PART V: REDUCING THE THICKNESS OF ROLLER POSTS

The reduction of the diameter of a roller post to fit a small hole in the roller is an operation which should not be undertaken without some understanding of how the roller should fit. When a roller is too tight on the roller post, it should not be forced. (See chapter "How to Match a Balance Staff," Figure 29).

The roller post may be reduced in thickness until it permits the roller to go about three-fourths of the way down this post on a loose fit. If the post is more than .05 mm the thickness of the sample staff, it may be reduced by turning with a graver until it is within .015 mm of the required diameter (measured at the base). Then it may be ground and polished to size and fit.

If an oilstone slip is used to reduce a roller post, it will not reach into the sharp-cornered base of this post. When the roller is pressed on, it will not reach the base of the balance hub, fitting as shown in Figure 25. This will prevent a sturdy support of the roller as well as prevent the pallet fork from properly engaging the rollers.

The shovels provide an excellent means of properly reducing the roller post. The shovel is placed under the roller post and the eccentric tailstock taper is turned until the flat surface is slanted to match the taper angle of the roller post as shown in Figure 23. The reduction and polishing of this part are done precisely as was described with the polishing of the balance pivots. The surface of the shovel is dressed with the file and the powders mixed in the same manner. The sharp edge of the shovel carries the abrasive into the corner of the roller post base as

FIG. 25. An oilstone slip will not reach into the corners of the roller post, thus preventing full seating of the roller.

well as the surface along its length, reducing it all equally. The tapered surface on one side and the tailstock roller on the other steady the shovel so that it maintains its angle. Slight upward pressure is all that is needed to give quick, efficient results.

Fitting Loose Rollers

Attempting to alter a roller to fit a staff whose roller post is thinner than required is contrary to the watch-making principle that no original part should be altered to fit a replacement.

When the roller post of the staff is so thin that the roller will not remain secure to it, it sometimes may be fastened to the staff if the adjustment to be made is slight. Staffs with posts too thin should not be used. Any method to make the roller fit must insure the concentricity of all parts of the roller with the staff center without altering or damaging the roller itself.

Some watchmakers pack the loose roller with a paste made from abrasive powders, the grade or coarseness of the grit depending upon the amount of space between the diameter of the roller post and the size of the roller hole. The roller, when placed over the roller post, displaces most of this paste. However, the fine particles of grit that remain become wedged between the walls of the roller hole and the post of the staff. This provides some security to the roller. The grade of abrasive powders used depends on how much space lies between the walls of the roller hole and the staff. The roller and staff may be cleaned with benzine; however, the abrasive grit, tightly wedged in the hole, will not become loose.

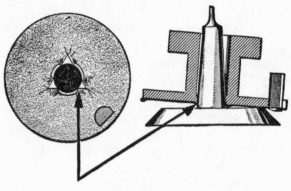

Fig. 26.

This method has the advantage that the character of the roller or staff has not changed and, if another staff need be fitted during the life of the watch, the roller has not changed in appearance or hole size.

Where the roller is very loose, the staff should not be used. Some watchmakers make a sleeve to fit over the roller post to build up its thickness to fit the hole in the roller. This is done by turning a piece of steel or brass wire to fit the roller post as well as a good staff should. Then a hole is drilled through its length. This hole is broached out to fit over the roller post of the staff. This little tube is then fitted half over the staff and the other half through the hole in the roller. The unit is then staked into place. Although this method works and is fairly satisfactory, it requires a great deal of skill, and the labor entailed could be extended toward making a completely new staff to dimensions.

A bad practice in fitting loose rollers utilizes the staking set equipped with the three-cornered punch which can be used to close an opening in the roller. By striking the hole with this punch, three burrs are raised, extending inward and thus reducing the opening in the roller hole. In the first place, this method damages the roller. Secondly, it provides a grip upon the roller post at one end only, leaving the rest of the roller clear of the staff post. The roller may then tilt to one side, upsetting its concentricity with the staff center and diminishing the validity of the safety roller. This is clearly shown in Figure 26. Avoid other methods of altering a roller to fit a thin staff that will damage the roller and impair its performance and accuracy.

PART VI: ADJUSTING A HAIRSPRING COLLET TO FIT THE STAFF

Hairspring collets are made of hardened brass, holed across to receive the hairspring and pin. The hole in the collet is chamfered on one end to facilitate its entry upon the hairspring shoulder of the staff. The collet is split to provide some elasticity in adapting itself to the staff's thickness. However, if this is forced upon a staff too thick for its use, it may spread excessively and cause its sides to touch the inner coil of the hairspring, spoiling its timing qualities as shown in Figure 27. This may also cause the collet to crack. When the collet is spread too much, it does not grasp the staff completely but becomes oval, gripping at about three points and providing an insecure seating at best.

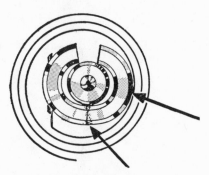

FIG. 27. Results of forcing a tight hairspring collet.

A hairspring shoulder should be properly dimensioned so that it does not require the collet to be spread. Staffs that gauge over .03 mm of the measurement of the original sample should be turned down with the graver and finished by grinding and polishing with the shovels prior to the staking of the balance. The manner in which a hairspring emerges from the collet and develops into its spirals is greatly responsible for the accuracy of its timekeeping values.

Fitting a Loose Collet to a Staff

When a collet is loose upon the staff, it may be closed up to certain limits. This limit is the closing of the collet so that the split in the collet is "closed" and no more space exists at this split. Any attempts to close the collet beyond this point to make it fit the staff will only result in damage to the collet and hairspring.

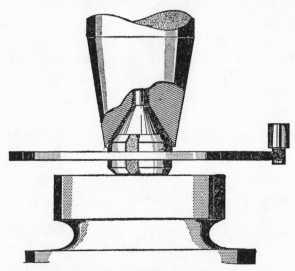

Fig. 28. A mouth-tapered punch may be used to contract the split in the collet.

A collet may be closed by placing it and the spring upon a flat steel surface such as a bench anvil or staking die stump. A taper-mouth punch is placed over the collet as shown in Figure 28. Care must be taken that the lips of this taper-mouth punch do not touch the hairspring. The punch chosen must fit these requirements. A slight tap of the hammer on this punch will close the collet to the limits stated above.

Some collets have the holes for the hairspring drilled below the center. In such a case, as in a breguet spring, the part of the collet giving the greatest clearance to the punch is made to face up to the mouth of the punch.

If a base with a series of graduated taper-mouth countersinks is available, the hairspring collet may be placed into the hole that will permit the edge of the collet to sit in the countersink without the flat edges touching the hairspring. Then a fairly wide flat punch is brought over the collet and tapped lightly. This also will close the collet concentrically as illustrated in Figure 29.

Screwplates have a series of such graduated countersinks and are used by many watchmakers as a base for this operation. When the collet is tapped, it may become stuck momentarily in the countersink. It is easily removed by pushing it out from the back of the hole in the countersink.

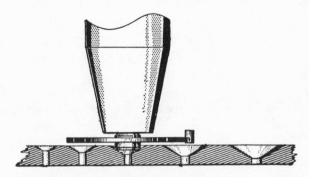

Fig. 29. A series of countersinks makes a handy accessory in collet-closing.

A slight smear of oil placed in the countersink before the operation will exclude this possibility.

Attempts to fit the collet after these methods fail will only damage the collet and possibly the spring. Where it is necessary to retain the staff and the collet is much too loose, it would be better if the collet were replaced for one that would fit.

Summary

1. The pivot lathe is used to polish or reduce the thickness of conical pivots.

2. The pivot must fit through the disc so that all its cylindrical portion may be polished.

3. The burnisher is used with a smear of light oil.

4. Pivot files do not require oil.

5. Arkansas oilstones or abrasive powders should not be used with pivot lathes.

6. Burnishers or jasper stones must be applied so that pivots remain cylindrical.

7. Oil is not used with a jasper stone.

8. The Jacot tool is superior for use in adjusting all types of pivots.

9. The pivot should be .02 mm higher than the top of the groove of the Jacot runner chosen.

10. Burnishers come with both right- and left-handed edges.

11. The disc-end holes are used to reduce or polish pivot ends.

12. Polishing shovels are used to obtain even, straight surfaces or shoulders and pivots.

13. Shovels are used in conjunction with an eccentric tailstock taper.

14. Abrasive or polishing powders are used as the reducing or polishing agent.

15. Spatulas mix the powders to the proper consistency.

16. Oilstone slips should not be used to reduce shoulders with square corners.

17. Rollers should not be marred or "nicked" to make them fit a loose staff.

18. Packing the roller hole with coarse abrasive grains may sometimes secure a roller to a loose staff.

19. Collets that are spread excessively may break or may spoil the hairspring.

20. Collets may be closed with a taper-mouth punch, fitting over the collet and not touching the hairspring.

QUESTIONS

1. What type of staff is selected when an exact duplicate is unavailable?
2. When a staff is too high, what factors determine where it should be made shorter?
3. How is the staff made shorter?
4. What tools are used to adjust the pivot thickness?
5. What precautions are used to keep the cylindrical portion of the pivot straight?
6. How can pivots be adjusted without the bench lathe?
7. How is the proper groove in the Jacot tool selected?
8. Why are Jasper stones or oilstones prohibited when using the Jacot tool?
9. How are pivot ends adjusted?
10. Explain the use of oil with (a) the oilstone, (b) Jasper stone, (c) burnisher and (d) pivot file.
11. When is a left-handed burnisher used? A right-handed one?
12. What are polishing shovels?
13. How are thick roller posts made thinner?
14. What are some rules in adjusting rollers to fit a loose staff?
15. What are the results of spreading a hairspring collet excessively?
16. How may collets be closed? How much?

X

HOW TO TRUE A BALANCE WHEEL

PART I: TRUING IN THE FLAT

Truing a balance is a necessity growing out of needs more important than the symmetrical appearance of the balance. A balance wheel that wobbles up and down needs more room and space to carry on its eccentric motion. Aside from the poor performance of the watch, the wild gyrations of such a wheel will cause it to scrape against other parts of the watch, causing possible stoppage. Balances "out of round" are also out of poise. This is best explained by the fact that a wheel that is not perfectly round has some parts of its circumference or wheel rim nearer to the center and other parts of its rim are farther away from the center. Since all parts of the wheel are not equally distant from the center, those sections farther from the center act as heavier portions. Those nearer to the center become the lighter sections of the balance. Conditions such as these prevent anything like close timing.

To true a balance, a few simple tools are used. The main piece of equipment used is a balance caliper. This is shown in Figure 1. The balance wheel is placed in the nibs of the balance caliper. The caliper is equipped with two sets of nibs: A and B. For truing, the balance is placed between the nibs marked B. The magnified illustrations show how the nibs B prevent the balance staff pivots from breaking. It will be noticed that the holes drilled down into the nibs are chamfered at the top so that the cones of the staff, rather than its pivots, rest against the nibs as shown by the arrow in inset B. The hole drilled across and through the nib is used to aid in observing the pivots and in removing any broken pivot that might become lodged in the holes of the nibs, as well as to aid in keeping the holes clear of dirt and grit.

The nibs A are used when the balance is to be poised and for the observation of train wheels for truth. However, for poising balances, the more reliable poising tool is to be preferred. When the nibs A are used, the caliper must not be tightened as this will only press on the pivot ends and break them as the pivot ends rest against the coned section of the nibs.

Balances generally assume certain patterns when they are bent. These

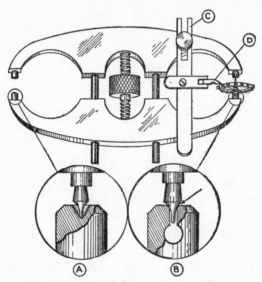

Fig. 1. A parallel balance truing caliper.

are caused by improper fittings of the balance staff, that is, the balance shoulders are either too tight or too loose. Other causes are staking the staff too tightly and inexperienced manipulation of the balance or the result of accidental bending. Careless removal of the old staff also distorts the wheel. In this chapter, the most common bends will be discussed.

To obtain skill in truing balances, it would be well for the beginner to take some old 18s balance wheels, staked to staffs. Bend these into the shapes shown in Figures 3, 4, 5, 10, 11, 14, and then follow the instructions for their repair. Also take these balances and bend them extremely out of shape so that the wheel resembles the letter "S" rather than a wheel. Also bend the balance so that one side of the rims points upward and the other downward.

The truing process starts, with the balance out of the caliper, by roughly bringing it back to shape as near to perfect as can be done with the aid of the fingers and the judgment of the eye. The balance is then placed into the calipers, the top side uppermost and in the nibs marked B. The index elevator C can be lowered or raised so that the indicator arm D is in position.

Figures 2A and 2B show how the indicator arm is placed into position. The indicator arm (2A) is placed over the top side of the balance so that just a bit of light can be seen through the space between the balance and the bottom of the indicator. In this particular case, an imaginary line (A)

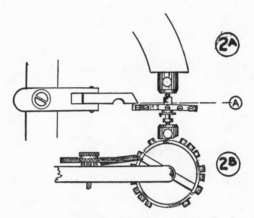

Fig. 2. Adjusting the index for truing in the flat.

drawn across, shows this balance to be flat until it reaches the fourth screw from the left, then it slightly and gradually becomes lower. The caliper indicator should overlap the balance rim slightly as shown in Figure 2B. Remember that the rim of the balance is under observation and not the screws.

Except in certain cases, practically all truing operations in the "flat" are done while the balance is in the caliper. After the index has been placed into position, the height of each arm of the balance must be made equal to one another. If they are not, bring one arm higher or lower so that both remain on a common plane and yet form right angles to the axis of the balance staff. This must be done before any other operation in the truing job can be attempted. The raising or lowering of a balance arm can be accomplished by placing the thumbnail underneath the low arm and raising it or by depressing the higher side with the index finger.

When the arms are equal, the caliper is tightened so that the balance may only be moved with the fingers. The indicator is again placed over one of the arms and the balance is observed through an eye loupe while the caliper is held so that the view is similar to Figure 2A. As soon as the slightest deviation is observed, it should be corrected.

A typical balance bend is shown in exaggerated form in Figure 3. This usually results from staking the staff too tightly, causing the balance arms to spread and forcing the rims to bend upward or downward. A simple method of bringing such a wheel back to true is shown. The thumbnail is placed against the underside of the split end of the balance wheel and the index finger, slightly in front of the thumb but resting on top of the middle

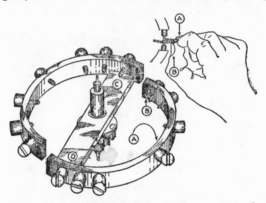

FIG. 3. In bends such as these, the fingers are used
to straighten the balance.

of the balance rim. A twisting motion is then used. The thumbnail presses the split end upward and the index finger, resting on position (A), acts as the fulcrum. Pressure is applied in the direction of the arrows. Sometimes, pressing the arm (C) upward and arm (D) downward will bring the balance back to true in the flat.

Figure 4 shows a balance where both ends are on level with the rest of the balance, but the middle of the rim sags. Truing such a balance is done as illustrated. The thumb is placed under the middle of the rim and the forefinger is placed on top of the split end of the balance. A twisting upward pressure is applied as shown by the arrows to bring this seemingly difficult bend to an easily solved job. The finger positions in all drawings should be closely observed if a successful effort is to be expected.

The most stubborn are the sharp bends which do not yield to finger manipulations. These bends are caused by mishaps at the bench such as dull screwdrivers slipping against the balance or dropping the balance. A form of such bend is shown in Figure 5. Here the correction is applied with the balance removed from the calipers after the exact spot of deviation has been determined. A mental note of the spot can be made by memoriz-

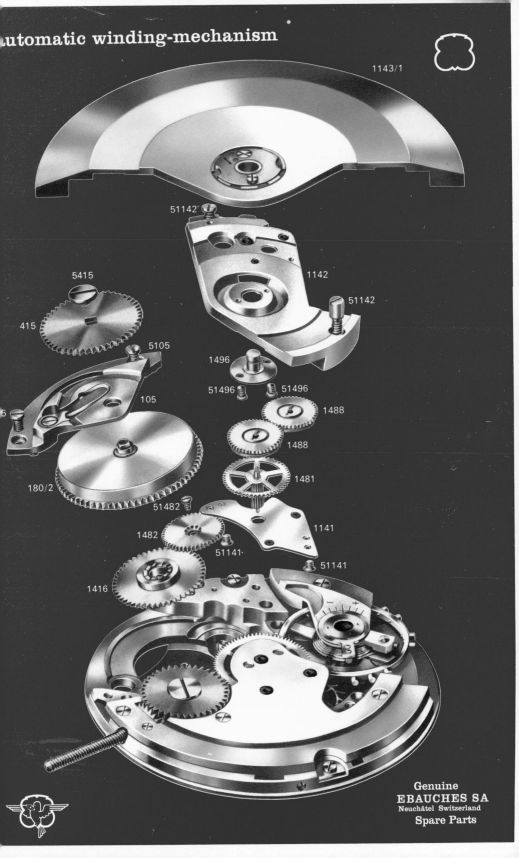

automatic winding-mechanism

1143/1

51142

1142

51142

5415

415

5105

1496

51496 51496

105

1488

1488

180/2

1481

51482

1141

1482

51141

51141

1416

Electronic movement with balance-spring

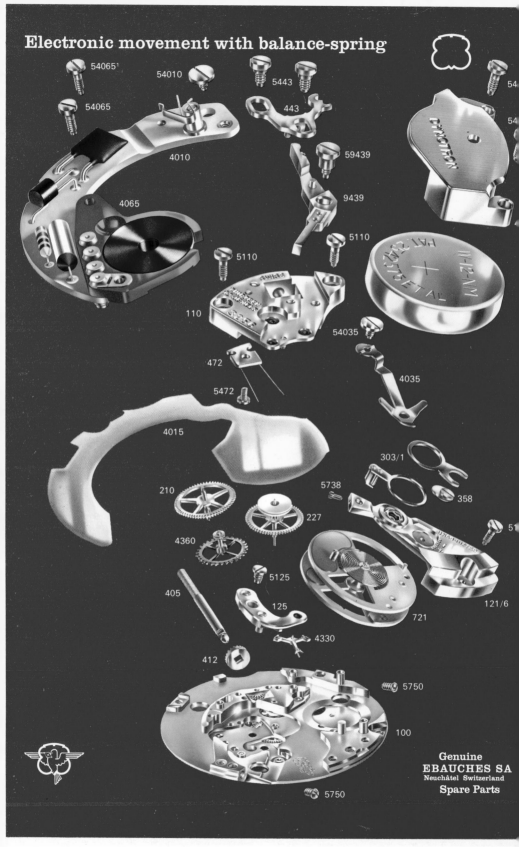

54065¹
54010
5443
443
54065
4010
4065
59439
9439
5110
5110
110
54035
472
4035
5472
4015
303/1
5738
358
210
227
4360
51
5125
405
125
721
4330
121/6
412
5750
100
5750

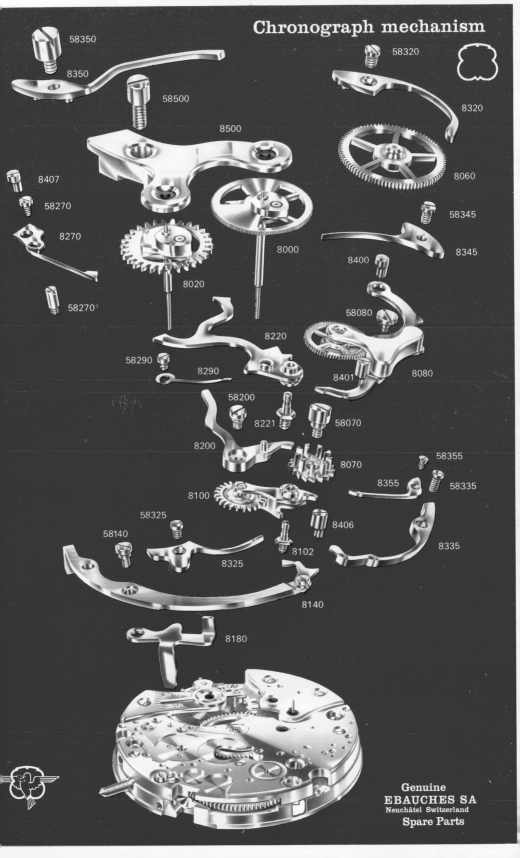

Chronograph mechanism

Genuine
EBAUCHES SA
Neuchâtel Switzerland
Spare Parts

ing the position by the screw numbers. In this case the bend takes place between the third and fourth screw from the arm. A pair of light parallel pliers whose jaws are lined with chamois or suède leather is used. The balance is gripped at the point of bend shown here as (A). The leather lining prevents the steel jaws from marring the balance rim. The balance is forced back into a level position while the pliers grasp the balance rim as shown in the inset drawing; the other parts may be bent back with the fingers or a flat-nosed tweezers also lined with leather. Bending must be done cautiously as it is far better to bend a little at a time than to overbend and strain the balance wheel.

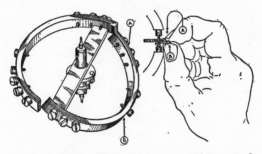

FIG. 4. Study these finger positions to aid in normalizing bends as these.

In truing a balance in the flat where the balance has more than one bend (Figure 6), the procedure is much the same. The bends nearest the arms must be treated first without regard to any other bend farther out on the rim of the balance wheel. The procedure is shown in the illustration in 6. Here the balance is in the caliper and the balance has been divided into four sections, each representing a different plane or bend. The indicator is again in position over the arm and an imaginary line has been drawn across the top of the balance in all four figures. This line is terminated at the Figure 1. When the balance has been trued in the flat, it must be parallel to this line (Figure 4). In Figure I, section A is flat; section B bends downward *and* its deviation is represented by the angle 2. Section C bends upward and its angle of deviation from the indicator line 1 is represented by angle 3. The last section, D, is seemingly only slightly bent but yet needs correction; its angle is noted at 4.

In this case, straightening is done again with the leather-lined pliers and tweezers. The first section to be corrected is B. This is done until it is perfectly level as in Figure 6, II. No further section should be treated until the parts nearest the arms are perfectly flat and true. To ignore this warning is to complicate the bends and to introduce new bends and kinks.

In II, Figure 6, section B has been straightened but the angles of sections C and D and angles 3 and 4 seem changed, appearing worse. This should not alarm the person truing the balance. He should continue truing the balance in the manner illustrated in Figure 5.

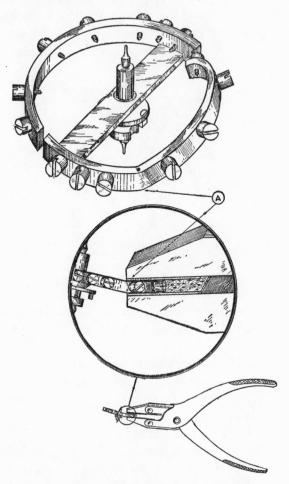

Fig. 5. Chamois-lined parallel pliers are recommended for straightening acute bends in the flat.

In III, section C and its angle have been eliminated by truing; this is shown completed in this figure although section D seems to have become worse; the truing process should complete the job shown in IV. Here the balance rim is perfectly parallel with the indicator arm. This is tested by rotating the balance and noting the amount of light and space between the indicator and the balance.

All illustrations up to now have been with the split or compensated type of balance. However, truing a solid balance can be equally annoying. A most common form of bend in solid balances is shown in Figure 7. One rim of the balance is slanted downward and the opposite end slants up. Shown with this drawing is a simple method of correction. If the index finger pushes the high end downward, the opposite end seesaws upward. In

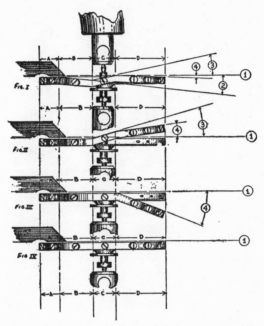

Fig. 6. Steps in normalizing a compound bend such as shown above.

this way a return to normal may be effected. Harsher bends and complex bends are treated as the bends illustrated with the split type of balance.

When bending the balance while it is in the caliper, the calipers must be tightened so that the nibs press against the cones of the balance staff pivots. This will prevent the balance from slipping out of the nibs and breaking the pivots.

Although these illustrations have shown exaggerated bends, they are pictured so that the conditions will be clearly visible. However, in practice, the slightest and smallest variation shown by the indicator must be corrected. This is only observed with the aid of an eye loupe and the indicator as close to the balance as possible. The balance should be turned with the fingers very slowly. Spinning the balance only results in a pleasant but illusionary effect of the balance being true. The speed of the balance blends all errors into a common line resembling a concentric wheel. Sometimes the balance will seem true in the calipers with very fine inspection but when placed in the watch, its motion is clearly eccentric. In this case suspicion should be centered on the balance pivots. These may not be straight or the sideshake in the jewels may be excessive.

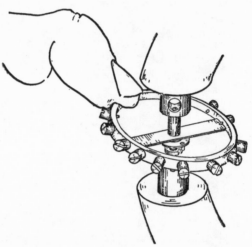

Fig. 7. Solid balance being straightened in the flat,

Summary

1. Balances that are out of round are out of poise.

2. Parts of the balance wheel farther away from the center act as heavier units than those nearer the center.

3. The balance wheel is placed in those nibs of the balance caliper which protect the pivots against injury.

4. Balances may become bent because (1) the staffs were driven on too tightly, (2) they were ill fitted, (3) they were accidentally bent, or (4) the staffs were carelessly removed.

5. The indicator arm of the caliper is adjusted just over the side and edge of the balance rim.

6. Both arms of the balance must be made equal in height.

7. Most bends in the balance may be rectified in the caliper with the use of the fingers.

8. Stubborn bends require the use of special pliers or tweezers or balance-truing wrenches.

9. Check for slight variations with the eye loupe.

10. The balance may be corrected while in the calipers when the jaws are tightened.

QUESTIONS

1. Why should a balance be trued?
2. What is the main tool used in correcting the balance in the flat?
3. How are the pivots protected against injury while in the caliper?
4. List some of the causes for bending of balance wheels.
5. What is the function of the indicator arm?
6. Describe how the fingers are used in straightening the balance wheel.
7. How would you straighten a balance wheel that resists the efforts of the fingers alone?
8. Explain fully how you would straighten a balance with more than one bend.

PART II: TRUING IN THE ROUND

Truing the balance wheel in the round is an operation that requires more skill and exactness than truing in the flat. Furthermore, the necessity for a perfectly round wheel is considered of greater technical importance than a perfectly flat wheel. The slightest deviation from a truly round wheel is easily detected in a watch and detracts considerably from the appearance of the balance while it is in motion. Such errors also prevent close timing and adjusting as explained earlier.

To start the truing operation in the round, the index is placed into the position shown in Figure 8, as close to the rim as possible and over the screw. Ignore the screws; we are truing the wheel rims and not the screws, which vary in height, thickness and weight. This position should permit as little light as possible to be seen between the rim and the index. (See Figure 8C.)

Starting at one of the arms, observe the amount of light between the rim and the index at this spot. Then turn the balance to the opposite side and compare the "light space" with that of the first arm. If the space between each arm is the same, it may be assumed that both arms are of equal length. Should one arm indicate more light space than the other, this arm is then shorter. The truing operation should begin by stretching the shorter arm so that it will equal its neighbor in length.

Illustrated in Figure 8, is a balance wheel where the arm has been divided into two parts, A and B. The length of each is clearly indicated by the spaces at C and D. D shows that the arm adjacent to the index at that point is shorter than the opposite arm. This may have been caused by a replacement staff that required extra spreading in order to become secure in the balance wheel hole, assuming an eccentric position in the balance

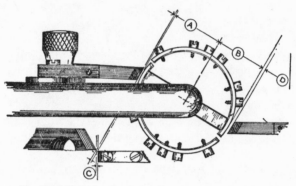

Fig. 8. Adjusting the index for truing in the round.

wheel so that the staff did not settle in the exact center of the hole. This caused one arm to become farther removed from the center than the other.

Stretching a Balance Arm

Figure 9 shows how the stretching operation is done. First, remove the roller and jewel and place the balance in the hole of a flat-faced stump or a punch inverted to act as a stump. The hole in the stump or punch should be just large enough to accommodate the hairspring shoulder of the balance staff snugly but not tight. Then a peening punch is brought over the balance arm as close to the staff as possible. The peening punch is tapped in a series of light blows which should stretch the balance arm. The result should be checked often in the caliper. Over-stretching is worse than none and should be guarded against. Stretching of the arms is done nearer to the staff so that the mark left by the peening punch is concealed when the roller is replaced.

When the arms have been made equal in length, the truing of the balance should proceed, concentrating on one half of the balance wheel, working outward from the arm toward the split end of the rim.

Two typical bends in the round of a balance are shown in Figures 10 and 11. In Figure 10, it can readily be seen that one of the arms has been bulged out in the middle and bent inward toward the split end. This can be brought close to normal by using the balance truing wrench, hooked as close to the split end as possible and pulling outward in the direction shown by the heavy arrow. The dotted line A, indicates the position the rim B must assume when it is "trued."

In Figure 11, the reverse of Figure 10 is shown and the correction may be accomplished simply by placing the soft tip of the thumb under and

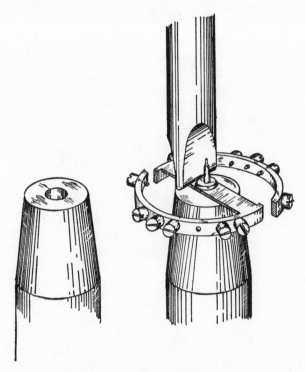

Fig. 9. The punch serving as a stump and the method of stretching a short arm by peening.

against the balance at "A" and the fingernail of the middle finger at "B" and then squeezing in the direction of the straight arrows, "A" and "B." Finger positions in all illustrations should be closely observed since experienced balance truers almost exclusively use the fingers to bring a balance to concentricity with remarkable speed and accuracy.

Figure 12, shown in two views, illustrates a balance bent outward immediately near one of the arms. The balance wrench shown, made of hard brass or nickel silver, is placed at the spot where the bend has been discovered. Moving the wrench in the circular direction of the arrow will return the rim to its correct position.

Figure 13 shows a bend in the opposite direction from that shown in

Figure 12. This is brought to the concentric position shown by the dotted lines with the wrench exerted in the direction of the arrow. The manipu-

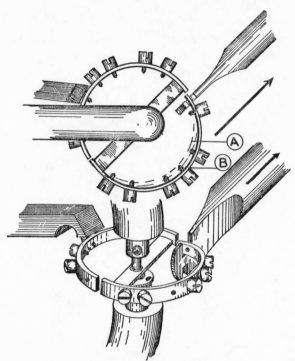

FIG. 10. Use of balance wrenches.

lation of the wrench and its position on the rim near the bends must be carefully observed so that using it will correct instead of create additional bends.

Where the balance has a number of bends (Figures 14, 15, 16), the

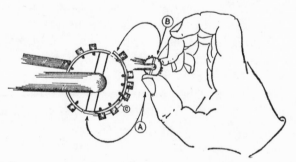

FIG. 11. The fingers are used to press a distorted balance
rim back into shape.

truing process is the same, starting again at the arms. The correction must be made at the exact spot that the first deviation is observed. After the initial error is corrected, the balance is turned slowly with the finger, past the index until the next deviation is noted and the correction again made. At times, correcting one part of the rim may apparently aggravate the condition of the rest of the rim. This should not alarm the manipulator; he

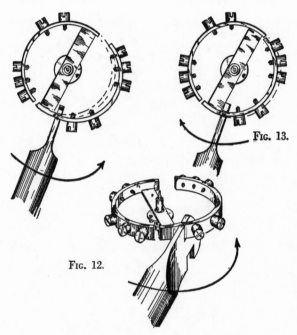

Fig. 13.

Fig. 12.

Another use of the truing wrench.

should continue the truing process until the balance is concentric without the slightest deviation discernible in the caliper and index. Observe Figures 14-15-16 closely.

Figure 16 shows the final bend being trued with a balance wrench made to straddle the screw. This is also shown in the close-up view in Figure 17. Such a tool has many advantages, especially where the bend is close to a screw. Removing the screws is inadvisable in most cases since the weakest part of the balance rim is the empty threaded holes in the balance rims. Balance wrenches are scarce; they are easily made, however, and should well repay any small efforts to produce them.

Sometimes, however, it is necessary to remove the screws in order to correct stubborn bends and often to polish or refinish a balance wheel. In

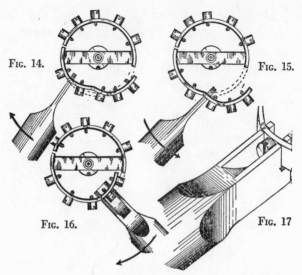

Steps in repairing a compound bend in the round.

Figure 17 shows how a slit wrench is used to straddle a screw.

such a case, removal of the screws must be carefully done so that the screws
are not marred or the screw slots disturbed. Some material or watch tool
catalogues show a number of effective tools for this purpose, such as
special screw-drivers or balance-screw holders. In removing the screws,

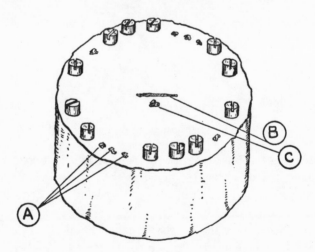

FIG. 18. A piece of soft pith may be used to hold balance
screws in an identifiable position.

especial care must be exercised not to mix them. They must be placed in a receptacle so that each screw may be replaced in the exact hole from which it was originally removed.

When the job of truing the balance in the round has been completed, the balance should then be rechecked for truth in the flat. This is necessary, since manipulating the rim of the balance in truing it in the round might have upset its truth in the flat. Any corrections needed should be made regardless how small and insignificant they might seem.

A simple device is shown in Figure 18. A piece of flat, round pithwood is used. A pencil mark or pencil line "B" is made to represent the arm of the balance. Another mark "C" represents the side of the arm on which the roller jewel is placed. When the screw nearest the arm is removed, it is inserted thread down into the soft pithwood nearest its simulated position

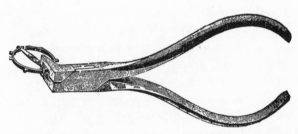

Fig. 19. A special pliers used to bend a balance arm without scratching the rim.

on the pithwood. Should the screw come from the arm on the roller jewel side of the balance, it is placed into the pithwood on the corresponding side. The next screw is placed near its natural partner. Should there be an empty screw hole between one screw and the next, the pencil point is depressed into the pithwood in that spot to represent the empty screw hole on the balance wheel. This is shown in the drawing at "A." In this manner, when the screws are replaced in the balance, the poise and temperature adjustments are practically undisturbed.

Figure 19 shows a balance arm pliers made to grasp the arms of the balance wheel without touching the rims. These are used to bend the arms in the flat positions or to hold the balance during manipulation.

Most of the truing operations should take place while the balance is in the calipers since quick observation and comparison are important.

Little can be said about truing monometallic or solid balance wheels in the round. When very slight variations are noted, these may be lessened sometimes by grasping the staff of the balance in the lathe chuck by the hairspring shoulder and twisting the balance wheel slightly around to a new position which may possibly be more favorable. If such a balance is brought to poise, the error is then minimized so far as practical purposes are concerned. In bad cases, the balance should be discarded since any attempts to round out such a balance usually induces the operator to use "botch" methods with corresponding results.

Selecting the calipers to be used calls for some advice. Parallel calipers are commendable; they permit the nibs to remain on the same vertical plane and axis as the staff. Also they can be locked, leaving greater freedom to the fingers for better manipulative power. The index of the caliper should always be a dull blue or black color so that the light does not reflect, giving a false impression as to the amount of light falling between index and balance rim. The black finish to the index may be renewed by dipping the index into oil and then allowing it to burn; the smoke deposit is fairly permanent and will serve satisfactorily. A new pair of calipers may be tested by closing them until both nibs touch and observing if they meet "head on." If they are in line, this will be the case; if they do not, the calipers should be rejected. Another test is to open the calipers fairly wide and try to twist one part from the other gently, feeling if there is excessive play or shake within the caliper steady posts. Excessive shake is indicative of a caliper that will not hold a "lined up" position.

In truing very small balances, a pair of tweezers may be used instead of the balance wrenches. Other accessories include flat tweezers lined with chamois leather, wooden cupped blocks to aid in bending the balance in the flat, and similar devices.

Summary

1. A balance that is out of round prevents close timing and adjusting.

2. Truing in the round is started in the balance caliper.

3. Both balance arms are of equal length if the light space between the rim of the balance wheel and the index is the same on both sides.

4. A balance arm can be stretched by the use of a peening punch.

5. A bend in the round of a balance can be corrected with the use of a balance truing wrench.

6. Fingers may be used in correcting a bend in the round.

7. If removal of the screws is necessary, they should be put aside carefully so that they can be replaced in their proper holes.

8. After truing the balance in the round, it should be checked for truth in the flat.

1. How does a balance out of round affect the timing of a watch?
2. Explain how you would determine whether both balance arms are of equal length.
3. State one reason for unequal balance arms.
4. How may the balance arms be equalized?
5. What tools would you use to correct bends in the round?
6. Under what circumstances may the fingers alone be used for straightening a bend in the round?
7. Explain briefly one method that may be used to avoid mixing balance screws that have been removed.
8. Why is it advisable to have the balance caliper index of dull color?
9. How would you determine whether a pair of calipers is in good condition?
10. What purpose do the cross holes in the calipers nibs serve?

XI

HOW TO POISE A BALANCE WHEEL

PART I: USE OF POISING TOOLS

Poising a balance is that operation in which the weights (screws) on the balance wheel are adjusted so that no section, screw or combination of screws acts as the heaviest part of the wheel.

After truing, the balance wheel should be poised. One of the greatest barriers to consistent close timing is lack of poise in the balance. Watches that may keep good time while lying in just one position will become erratic timekeepers when the watch is placed upright in the hanging (pendant) positions. By timing the watch in the various positions, an experienced watchmaker could accurately determine the exact spot on the balance which is heaviest and thus is causing the variation. But such a procedure would take much time and effort. It would be simpler to take the balance after it has been trued in the flat and round and place it on a balancing block or "poising tool" (Figure 1) and make all corrections.

All poising tools are made up principally of two parallel sharp edges. The balance is placed so that the cylindrical portion of the pivot-ends rests on these edges. When this is done, the heaviest part of the balance will roll down to the lowest portion of the poising tool, just as a bicycle wheel, spun freely off the ground, eventually rolls to a stop with the valve, its heaviest part, resting at the very bottom of the wheel.

While the truing calipers are used occasionally to poise the balance, the more sensitive poising tool is preferred. The superiority of this tool is proved when a balance which indicates apparent poise in the calipers will occasionally manifest errors when placed upon the poising tool.

The poising tool shown is made of heavy brass, bronze, or nickel silver. The use of steel is kept to a minimum to diminish the possibility of magnetic influence. The jaws "A" are made of polished agate ground to a sharp, strong, straight edge. These are set into blocks "B" and "B¹." Screw "C" is used to activate the lead screw "L" which moves block "B¹" back and forth along the grooved table to accommodate balances of all sizes

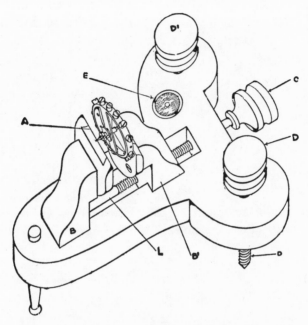

Fig. 1. A balance on the jaws of a balance poising tool.

and staff lengths. Screws "D" and "D¹" raise or lower each side of the tool so that the jaws will be perfectly horizontal, preventing the balance from rolling off the agate jaws. A helpful aid in this operation is the spirit level "E." The screws "D" and "D¹" are adjusted until the bubble is centered in the level. The level is not an absolute necessity since the balance itself serves this purpose. Effective poising tools may be made with hardened polished steel jaws but these must be checked constantly for magnetism which would influence the balance wheel and deceive the adjuster.

Before placing the balance on the poising tool, the balance pivots must be absolutely round and straight. Should the pivots be bent, oval, or have a slight "flat" spot on them (Figure 2), the balance will rock on these spots and it will be impossible to poise the balance. The pivots must also be clean and polished. The agate jaws should be clean and dry. Some watchmakers clean the agate jaws by running a clean piece of pithwood across them before setting the balance pivots on the tool.

The balance is placed on the agate jaws so that only the cylindrical portions of the balance pivots rests on them. This is shown in Figure 3. The poising operation consists of removing weight from a heavy screw or adding weights (timing washers) to the opposite screw. In this way an effect of poise may be obtained. However, there are certain definite procedures which must be followed.

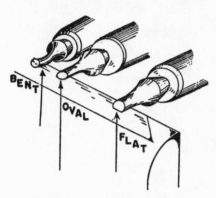

FIG. 2. Three types of pivots which may prevent accurate poising.

A knowledge should be had of the screws on the balance and their purpose. Figure 4 shows a balance set in a front view in its position on the poising tool. Screws B^1, B^2, B^3, B^4, B^5 and the BR screws, directly opposite them, are regular balance screws which give the balance a certain mass and weight. These screws are the only ones to be manipulated to effect the poise. The four screws, MA and MR, are

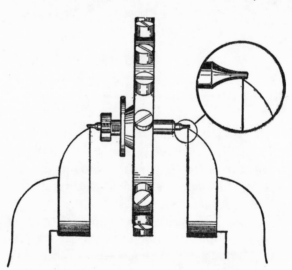

FIG. 3. Pivots must rest on the jaws of the poising tool by their cylindrical ends only.

meantime screws and are generally made of a heavy metal such as gold or platinum. These screws, used only in the better grade watches, regulate the watch to within tolerances of a very few seconds a day. These should not be removed or turned unless their purpose is completely understood.

To aid in identifying each screw on the balance and to recognize the difference from its diametrically opposite partner, all balance screws on the side of the balance in which the roller jewel is set will be called "BR" screws; those on the other side of the rim are called simply "B" screws. The screws get their numbers from their position on the balance with the roller jewel facing front and starting at one of the arms, counting clockwise. In Figure 4, the heavy screw is "B^2." Its opposite screw, acting as the lightest screw, is "BR^2."

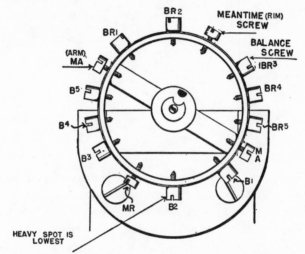

Fig. 4. A method of identifying the different balance screws.

When meantime screws are used, they are known by pairs, such as the meantime arm screws (MA) or the meantime rim screws (MR). Again, these are not to be touched in the poising operation. To turn these screws but a half turn will effect the timekeeping values by as much as a minute a day. Turning these screws is resorted to by better watchmakers who do not wish to move the regulator from its neutral position on the bridge, both for the sake of appearance and definite technical advantages.

Before removing weight from the heavy screw or adding weight to the lighter (opposite) screw, an inventory of the watch's past timekeeping record should be had. If the watch previously ran slow, it would be safe to remove weight from the heavy screw as this procedure will make the watch gain some time. If the watch previously ran fast, a weight (washer) might be added to the opposite screw, equalizing both errors at one time. However, the adding or removing of weight from the screws should not be indiscriminate. This will upset the regulation of the watch. When all the above factors have been determined, the balance is placed on the tool and the heavy spot is noted. Should it be decided to deduct weight from the balance, the screw must be removed.

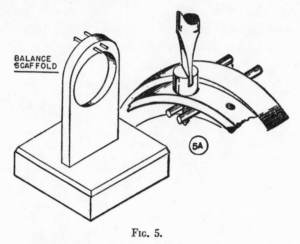

BALANCE SCAFFOLD

5A

Fig. 5.

Removing the screw also calls for some care and thought. The balance screw must not be marred during this operation. Neither should the balance be roughly handled so that the previous truing operation is disturbed. In order to remove the screw without disturbing the truth of the balance, it is placed on the balance scaffold shown in Figure 5. The threads of the

BALANCE SCREW REMOVER

Fig. 6.

screw are straddled by the two pins and the rim resting on these two pins. These scaffolds are equipped with two sets of pins, one pair farther apart to accommodate the threads of thicker balance screws. The screwdriver, one that is very sharp, gradually tapered and ground to fit the screwslot perfectly, is brought over the screw as shown in Figure 5a, and the screw removed.

In cheaper balances where the screws are not slotted, the balance screw remover may be used. This is shown in Figure 6. The prongs or jaws of this tool comes in various sizes of hollowed openings numbered 1 to 4 to accommodate the assorted sizes of balance screws. Using but one size tool for all size screws will only mar the screw unless it happens to be the size of the jaw-opening used.

Part II: Removing Weight from the Screws

There are many methods used to remove weight from the screws. Four of the most commonly used are shown pictured in Figures 7, 8, 9, 10. Figure 7 shows how a screw is undercut to remove weight. This method

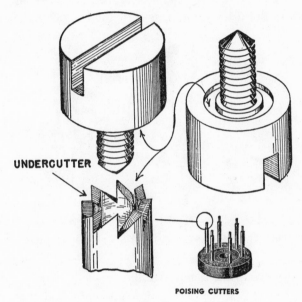

Fig. 7.

has the advantage that the screw is not changed in outward appearance. For this particular operation, the undercutters are used. The undercutters shown come in a set of six hollowed rose cutters. The sizes are arranged with respective hollow sections to accommodate the threads of the screws.

The screw to be undercut is placed thread down into a hollow tube cutter that fits around the screw threads closely but not tightly and whose outer diameter is less than that of the balance screw. The screwdriver is then placed into the screwslot of the screw resting on the undercutter and turned with a slight downward pressure. This will mill out an undercut as shown in Figure 7.

Figure 8 shows a fine screwhead file removing weight by cutting the screwslot deeper and wider. This is done also while the balance is on the scaffold. Special balance screw-slotting saws are also available for this same purpose. Here caution must be exercised so as not to remove weight excessively and also not to weaken or mar the screw.

Figure 9 illustrates a fine pivot drill countersinking the head of the screw. This is often seen on the tops of slotless screws used in the less expensive Swiss watches. When a slotless screw is encountered which must be made lighter, it might be more advisable to remove weight by simply using a screw-slotting saw to slot the screw. This will improve the screw and remove weight at the same time. The worst method, that of filing the screw diagonally from the bottom, is a method employed in Swiss watch

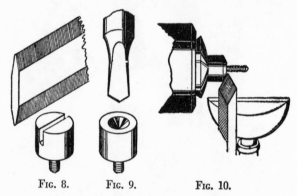

FIG. 8.   FIG. 9.   FIG. 10.

Other methods of reducing the weight of a balance screw.

factories on very cheap watches. Its only excuse is that it is done quickly where production and low price are paramount. Although this is a decidedly poor practice, it only stresses the importance of balance poise even in watches of the poorest construction.

Figure 10 shows a method used by many good watchmakers. The bottom edge of the balance screw is beveled or rounded off with a well-finished graver while the screwhead is held in a suited chuck in the lathe. This has the advantage of preserving the appearance of the screw. It also allows the screw to be fastened snug into the rim without the flat portion pressing against the rim and possibly distorting it. Other methods used to remove weight from the screws are to polish the tops of the screwheads in a screwhead polishing tool and to strip or slightly turn down the thickness of the screwhead on the lathe. Both are acceptable as good practices.

The amount or weight of the metal removed is left to the judgment and experience of the watchmaker. However, it is always best to remove less than to overcompensate. Usually a balance heavily out of poise drops to its heaviest spot rapidly and swings in wide arcs until it stops. A balance slightly out of poise rolls to its heavy spot rather leisurely. This generally should serve as a guide to the amount of weight to be added or subtracted.

Before replacing the screw, it must be cleaned. Another important factor is to avoid handling the balance with the fingers, as the heat from the fingers will expand that portion of the balance held. This expansion will show up as an error in poise, especially in bi-metallic split balances. Replace the screw by turning it snugly up to the balance rim. If the same screw again rolls to the bottom, repeat the process by removing weight. An indication that you are proceeding correctly is the gradual diminishing of the speed with which the heavy spot rolls and then rocks to the bottom. Should the heavy spot shift to another screw close on the same side of the balance, it indicates that you are making progress. Let us say that the heavy spot has shifted to "B3." Then, you must assume that screw "B3," now at the bottom, is the heaviest screw. Repeat the same procedure performed with "B2," guided, of course, by your judgment as to the weight needed to equalize the poise. Should the heavy spot settle at a place between two screws, then both these screws are to be adjusted.

Should the heavy spot settle at one of the meantime screws, then certain observations must be made. Notice if the opposite meantime screw is extending out of the balance rim the same distance as the meantime screw in question. This is noted by the number of threads visible between the rim and the bottom of the screwhead. If the heavy screw is extended farther, this may be turned in to equalize the opposite meantime screw or the lighter screw turned out accordingly. Either operation might equalize the poise here. Should both timing screws be extended equally, then they should not be touched at all but the alteration be made by removing weight from one balance screw on each side of the heavy meantime screw, or adding washers to the opposite screws bordering the light meantime screw.

Should the heavy spot suddenly shift to the opposite side after removing weight, too much weight has been removed. This is a bad step in the direction sought. In such cases, weight must be added to the original screw manipulated. This is done by adding a washer as shown in Figure 11.

Continue the poising operation, using all precautions listed before. The balance should be turned weakly with a camel's-hair brush. If it rolls slowly to a stop, repeat this process and note the point at which it does stop. If the balance continually stops with the same screw at the bottom, that screw is still heavy and must be adjusted. If the balance stops at a random screw or not at any repeated place, the balance might be in poise.

To test this further, tap the table of the poising tool with the back of the tweezers or stroke the screw threads of the poising table legs. This will set up a vibration and set the balance into motion, rolling slightly along the jaws of the poising tool. If there is still a predominant heavy point, that spot will then roll to the bottom or remain motionless. If the balance continues to roll and stop at no particular repeated point, the balance is assumed to be in poise and this operation is completed.

Sometimes, drafts in the room will cause the balance to roll on the delicate jaws of the poising tool and confound an accurate observation. To

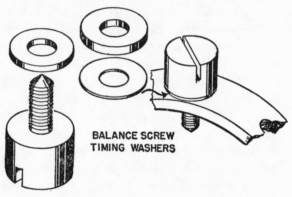

BALANCE SCREW
TIMING WASHERS

Fig. 11.

overcome this, a large, clear glass tumbler might be used to cover the poising tool while the balance is on the jaws; now the poise can be observed undisturbed.

When, during the poising operation, it has been found necessary to add weight to the balance, timing washers can be used. Shown in Figure 11 are three different thicknesses of timing washers. Timing washers come in sets and these are numbered in the sets so that their value in minutes a day added to the timekeeping is noted. When regulating a watch, they are used in pairs: one washer to one screw and its pair on the screw diametrically opposite. When poising a balance, only one washer on the light screw is used. The thickness of the washer chosen determines the weight to be added to affect poise. When the difference in thickness cannot be accurately observed, then the same diameter washer with a large time value is the heavier one.

Timing washers used must be of the same diameter as the screw or slightly smaller. If a larger washer is used, it may protrude over the edges of the balance wheel and catch on the center wheel or the pallet bridge, thus stopping the watch. (See Figure 12.) Likewise, one should not stuff the balance with washers or use more than one washer to a screw. To do so would, in effect, make the screw longer and cause the screwhead to scrape or catch against some part of the watch. To the apprentice, this cause is one that eludes observation and results in many puzzling moments as to the stoppage of the watch.

Washers must be chosen and inserted so that they are not distinctly visible, blending with the screw for a pleasant appearance.

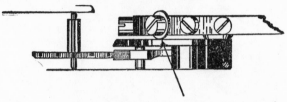

Fig. 12. An excessively large timing washer may stop the balance.

Poising a balance is necessary both to the largest and smallest balances. Solid and split balances are poised in the same manner with the exception that the split balance is more delicate and must be treated accordingly. Any adjustments to the balance must respect its state of poise and truth and nothing should be done to disturb this state.

To achieve skill in poising a balance, practice is essential. Old 18-size balances are good for these exercises. With them, you may add or remove weights and generally experiment to your satisfaction and edification.

Poising Screwless Balances

Modern horology has advanced sufficiently to enable a balance to be manufactured which does not require the rim to be loaded with screws in order to bring it to time or poise. Such a balance, produced by precise manufacturing methods, offers many advantages. It does not catch lint, is not weakened by holes in its rim which must be threaded and is free from air turbulence which the protruding screws cause. However, since the rim is without screws, weight cannot be added. If poising is required, weight must be removed. This is done as shown in Figure 13. A small pivot drill is inserted in a screwdriver handle and small countersinks exactly at the heavy spot *underneath* the balance are made. The watch is then brought to time (made to run slower to compensate for the weight removed) either by the regulator or by letting out some of the hairspring through the stud.

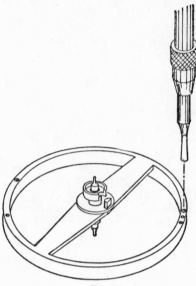

Fig. 13.

However, if the repair methods shown in this book are followed, these sturdy balances will not require poising or truing.

Summary

1. Consistent accurate timing may be prevented by lack of poise in the balance.

2. All poising tools are made up of two parallel sharp edges. Pivot ends rest on these edges.

3. The heaviest part of the balance wheel is always at the bottom on the poising tool when the balance is out of poise.

4. Before poising, the balance pivots must be round and straight, and polished; jaws of the tool must be clean and dry.

5. Poising consists of removing or adding weights to the proper part of the balance.

6. Removing weight makes the watch go faster.

7. Adding weight makes the watch go slower.

8. To remove slotted balance screws a finely sharpened screwdriver is used.

9. A balance screw remover is used if the balance screws are unslotted.

10. Common methods of removing weight are: (a) Undercutting, by the use of a hollow rose cutter of the proper size to fit around the screw threads; (b) cutting the screwslot deeper and wider by the use of a fine

screwhead file or a screw-slotting saw; (c) countersinking the head of the slotless screw by use of a diamond-pointed drill; (d) rounding off the bottom edge of the screw by means of a well-finished lathe graver; (e) polishing the tops of the screws in a screwhead polishing tool; (f) "stripping" the screwhead on the lathe with a well-finished graver.

11. The amount of weight to be removed can be determined by the speed with which the balance wheel rocks back and forth on the poising tool.

12. The removal of too much weight may be compensated for by the addition of timing washers.

13. When poise is affected, a washer of the proper thickness is added to the light screw.

14. Timing washers must be the same diameter as the screw for which it is used.

Questions

1. Why may a watch keep good time in the flat position yet be unreliable while in the hanging position?
2. Explain what is meant by poise in the balance.
3. What is a poising tool? How should a balance be placed on the tool?
4. Why must the pivots be round and straight before poising the balance?
5. Explain briefly why the heaviest part of the balance will fall to the bottom of the poising tool.
6. Of what does the poising operation consist?
7. What is a meantime screw? What purpose does it serve?
8. Explain a method of memorizing the various balance screws.
9. Under what circumstances is weight added to a screw? Subtracted?
10. What is a balance scaffold?
11. How are slotted balance screws removed?
12. Explain how unslotted balance screws are removed.
13. How can you determine which screw on the balance wheel is too heavy?
14. What tool is used for undercutting a screw?
15. Name two tools which could be used to cut the screw-slot deeper and wider.
16. How would you decrease the weight of an unslotted screw?
17. How may the lathe be used to reduce the weight of a screw?
18. How can you tell whether too much weight has been removed from the screw? How would you remedy this situation?
19. Explain briefly how you would determine the size of the timing washer to be used with a screw.
20. What may be the result of using a timing washer too large in diameter?

XII

REPLACING A PALLET JEWEL

A pallet jewel should be replaced if chipped or broken and, of course, if it is missing. A chipped pallet jewel (Figure 1) should be replaced with a new one because it detracts from the impulse it must impart to the

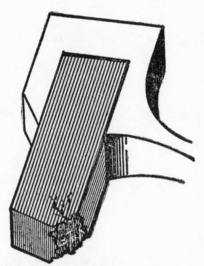

Fig. 1. A chipped pallet jewel.

balance. It may scratch the top of the escape wheel teeth and it may prevent a correct locking of the escape wheel tooth upon the other pallet jewel.

Removing a Pallet Jewel

Pallet jewels are held in the slots of the pallet by small bits of shellac (melted) to hold the jewel to the pallet. (Figure 2.) To loosen the old jewel, its binder of shellac will have to be warmed to its melting point

204

without overheating the pallet. This is done by first heating a thick piece of brass or copper called a "pallet warmer" over the alcohol lamp and then placing the pallet, bottom side up, upon it. The heat from the metal will warm the pallet sufficiently to melt the shellac holding the jewel.

An efficient pallet warmer is pictured in Figure 3, with dimensions supplied to aid in its construction. This tool is divided into two connected sections. These sections are separated so that only one pallet jewel may be heated without disturbing the shellac on the other. The hole is drilled to prevent the conduction of heat from one wing of the tool to the other. Two pointed steel legs (A) are inserted to keep the tool from scorching the bench.

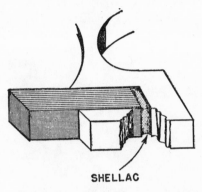

SHELLAC

Fig. 2. Part of pallet frame cut away to show where shellac is placed to hold jewel.

Similar tools are sold commercially and perform their tasks fairly well. Unfortunately, many of these are too thin and lose heat too rapidly, necessitating frequent reheating. Some watchmakers simply use a thick copper coin with a handle drilled into it.

The tool should be heated on one wing only. The pallet must not be on the tool while it is being heated over the flame. The tool is sufficiently hot

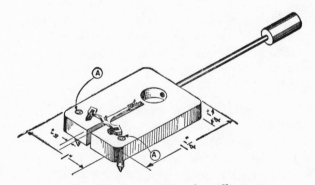

Fig. 3. Details on construction of a pallet warmer.

when the top of the steel legs "A" show the slightest change in color. The pallet is then placed, bottom side up, on the tool. A sturdy pair of tweezers grasp the pallet fork close to the arbor while a pointed brass tapered pin is used to push the jewel out from the rear as shown in Figure 4.

When the old jewel has been removed, all traces of shellac should be removed from the slot of the pallet. This is done by dipping a clean, pointed piece of pegwood in alcohol and scraping all parts of the empty

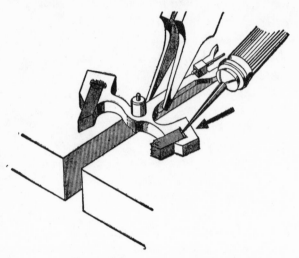

Fig. 4. Removing a pallet jewel.

pallet slot and adjacent areas which may show traces of shellac. The pallet is then cleaned in benzine and thoroughly dried.

Selecting a New Pallet Jewel

In replacing a pallet jewel, a knowledge of escapement theory and practice is indeed helpful. However, a pallet jewel may be replaced with satisfactory results by following a few simple rules. Although both pallet jewels may look alike to the untrained eye, they are different. They are easily distinguished by the experienced watchmaker because of the difference in the angles of the front ends of these jewels.

Pallet jewels are known by their position in the pallet. In practically all lever watches, the escape wheel turns clockwise, noted from the "dial-down," movement-side-up position. The pallet itself has two wings or arms. The jewels extend from each of these. The escape wheel teeth, in their clockwise motion, first drop on one of these jewels and then enter in between the pallet jewels, making their exit from the pallet by dropping off the opposite jewel. Therefore the jewels are called *entrance* and *exit* jewels, respectively. The entrance jewel is also called the right jewel, and the exit jewel, or stone, is sometimes referred to as the left jewel or stone. The terms "right" and "left" are applied to the jewels because most watch-

makers adjust the pallet looking at the pallet and escape wheel from the
balance towards the escape wheel.

In Figure 5, the pallet is pictured with an escape tooth locked against
the entrance (or right) pallet jewel. The jewels are also shown removed
from the pallet and their differences indicated by the comparative angles
of each.

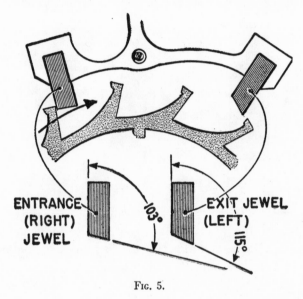

ENTRANCE (RIGHT) JEWEL   103°   EXIT JEWEL (LEFT)   115°

Fig. 5.

Where the sample is present, its thickness may be noted. The new jewel
should be thick enough to fit the pallet slot snugly. A jewel that is too
thin will not give the proper escapement action. This may also cause the
jewel to shift in its slot and introduce other faults detrimental to efficient
timekeeping.

If a replacement is needed for the entrance pallet, it should be one whose
lifting angle is not as great as the lift angle of the exit stone. Conversely,
an exit jewel when being replaced should have a greater angle than the
entrance jewel. Figure 6 concerns itself with the methods used in matching
the jewels for similarity of lift angles. In this picture, A^1 is the sample
jewel. A^2 is the replacement. To test it for lift surface, the replacement is
placed end on end to the sample jewel but with the jewels reversed on one
another so that their sides attempt to make a straight line. If the angles
are exact, the result will be a continuous straight line as in A^1 and A^2.
B^1 and B^2 show that the replacement B^2 has too great an angle and therefore
is unsuitable.

The length of a pallet jewel may be matched by placing them side by side so that their entrance corners or short sides touch each other as shown in C^1, C^2, Figure 6. The length of the replacement jewel should not exceed

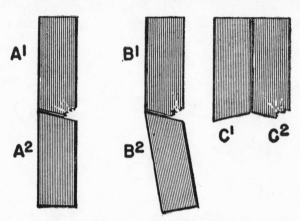

FIG. 6. Methods of matching pallet jewels.

that of the sample. This precaution is noted because a jewel that is too long causes the escape wheel to lock too deeply upon the pallet, taking motion from the balance and affecting the lock upon the opposite pallet as well. This seriously prevents accurate timekeeping.

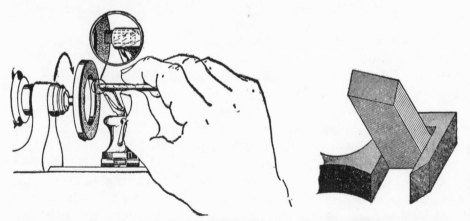

FIG. 7. A diamond charged lap may be used to shorten a long pallet jewel.

FIG. 8. Initial step in resetting a snug-fit pallet jewel.

Should a jewel be chosen with all dimensions correct except that it's too long, it may be shortened to an acceptable length. This is done by drilling a fine hole into the end of a piece of pegwood and placing the jewel, lift surface inside, so that the excess length is the only part protruding from the pegwood. The jewel is shellacked into place, and the protruding end is held against a diamond-charged lap or wheel similar to the kind currently used to sharpen carboloy gravers. This operation is shown in Figure 7.

When the jewel has been reduced to the proper size, it should be cleaned with alcohol after removing it from pegwood and readied for insertion into the pallet.

The pallet warmer is heated again and then the pallet is placed (bottom up) on the warmer. Caution must be observed so that the warmer is not

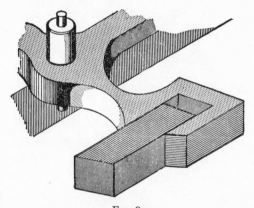

FIG. 9.

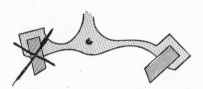

FIG. 10. A common error made by students is to reverse the pallet jewel as shown here.

overheated. Overheating this tool will only burn the shellac, causing it to crumble into dust as well as to discolor the pallet.

The jewel is inserted in the slot by grasping its sides with a pair of strong tweezers and bracing the pallet by holding it down with a brass pin. The jewel is then edged in by tilting the jewel and edging it, back end in, into the pallet slot as shown in Figure 8. The jewel is then pushed or tilted down flat, even with the surface of the pallet. The jewel should not be pushed in all the way but left just a bit short for trial observation as shown in Figure 9.

A common error made by apprentices is to insert the pallet jewel in so that its angle is reversed as shown in Figure 10. Of course, such an arrangement will not work.

The jewels must be placed in the pallet so that their inclined lifting surfaces face the oncoming escape wheel teeth. (Figure 5). The lift surface, therefore, must face counterclockwise when the pallet is in the watch. When the pallet is bottom side up, as it is in these illustrations showing the pallet on the warmer, the jewel is inserted so that the lift angle faces clockwise.

To secure the jewel to the pallet, shellac is applied to the area on the rear edge of the jewel and to the slot. Shellac for this purpose is made of the same substance but comes in two forms. One type is flaked shellac and is composed of flat chips. The other is called "shredded" shellac and resembles small threads of various lengths and thicknesses. Shredded shellac is made by heating a stick of shellac and sticking a piece of peg-wood into the soft, melted mass and drawing out small strings. These cool immediately and may be broken off and placed into a phial for further use.

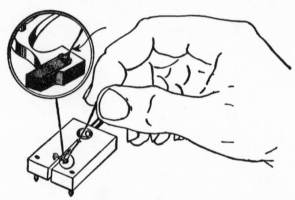

Fig. 11. Applying shredded shellac to pallet on the warmer.

Only a little shellac should be used. An amount as large as a typewriter period is sufficient to secure any size pallet. This small bit is placed over the junction of the pallet jewel and the rear of the slot. The pallet, when placed over the warm pallet warmer, will take enough heat from this tool to melt the shellac upon it. The melted shellac is held in place by capillary attraction as shown in Figure 2.

Shredded shellac is applied as in Figure 11. Here the warmed pallet will melt the shellac upon contact. The thin thread of shellac is held between the index finger and thumb. It is pointed at the junction and melts immediately. When sufficient shellac has been drawn into the pallet, the remaining part of the thread is withdrawn. Thus the amount of shellac may be regulated. Excess shellac is removed with the tip of a pointed piece of pegwood while the pallet is on the warmer. After the jewel is secure, the pallet is removed from the warmer, cleaned and placed into the movement and tried for escapement action.

Before proceeding further, it would be well to study Figure 12, showing the pallet jewel and the proper nomenclature of each part. This applies to both right and left jewels. Both have a locking surface, entrance corner, lifting surface (impulse face), let-off corner or edge, inside surface, top, bottom and backs.

This knowledge will aid us in understanding some of the escapement tests. One of these tests determines the correct length or position of the new pallet jewel.

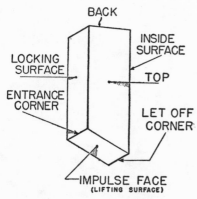

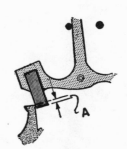

FIG. 12. Pallet jewel nomenclature.

FIG. 13.

This is done by placing the pallet in the watch movement so that the pallet fork rests against the banking pin on the side that has just received the replacement jewel as shown in Figure 13. Observe the lock at A.

"Lock" is measured from the entrance corner of the pallet jewel to the point on the locking surface where the escape tooth comes to rest. In a well-adjusted escapement, the lock should approximate ⅓ the thickness of the pallet jewel, as shown in Figure 14.

If the pallet jewel has less than this amount of lock, it may be extended. A pallet jewel that is too short will permit the escape tooth to enter directly upon the jewel's lifting surface. This will cause a "tripping" action, similar to an alarm clock hammer going off. This is shown in Figure 15. Here the pallet fork is resting against the banking pin at A and the escape tooth is upon the lifting surface of the pallet jewel at B. When this tooth drops off, the following tooth will fall upon the lift surface of the opposite jewel and again repeat itself on the entrance jewel. This will continue as long as there is power to drive the movement. Of course, no watch can work satisfactorily under such conditions.

Should the lock exceed the specifications shown in Figure 14, the jewel may be pushed back into the slot. This is done by heating the pallet warmer again and placing the pallet on it as in Figure 3. The tweezer is placed directly behind the slot of the jewel and the jewel pushed back the desired distance with a pointed piece of pegwood or a flattened brass pin pressed against the impulse face of the jewel. No additional application of shellac is necessary since the shellac applied originally is sufficient. This softens when the pallet is placed on the pallet warmer. When the jewel is pushed into the slot, the shellac will assume a position in the slot consistent with the space afforded to it and will harden as soon as it becomes cool.

Before manipulating the jewel or making any change, the following tests should be made:

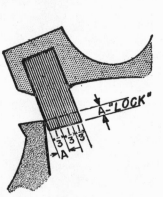
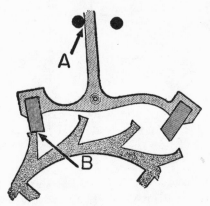

FIG. 14.   FIG. 15. Pallet jewel too short.

With the pallet resting against the banking pin, the mainspring is wound a few turns. Then a pointed piece of pegwood is placed in the pallet fork slot. The pallet is moved away from the banking pin with a very slow movement, permitting careful observation, until the escape tooth leaves the locking surface of the pallet jewel and enters upon the lifting surface of the jewel.

Still controlling the pallet and moving it slowly, observe the escape tooth as it advances along the pallet jewel until it drops off the let-off corner.

If the escape tooth refuses to drop off the jewel, it is because the pallet fork has already reached the other banking pin and cannot advance further to allow passage of the escape tooth. This is an indication that the new jewel is out too far and must be pushed in. The above condition is shown in Figure 16.

In this illustration, the escape tooth A pushes the entrance jewel upward and, of course, the fork travels toward the right. The escape tooth tries to drop off this jewel. Escape tooth C is ready to fall upon the exit jewel's locking surface, but the fork has already reached the opposite banking pin at B. The escapement is jammed at this point and the watch will not run.

We must always observe the rule that no part of the watch should be altered to suit a replacement part. Therefore the banking pins should not be moved without a prior knowledge of escapement theory and practice.

In order to permit the escape tooth to advance so that tooth A, Figure 16, may drop off and tooth C may lock upon the exit jewel, the new jewel will have to be pushed further into the slot until it meets the requirements of Figure 14.

A well-adjusted and -fitted pallet jewel should permit the escape tooth to drop off so that the opposite jewel receives an escape tooth on its locking

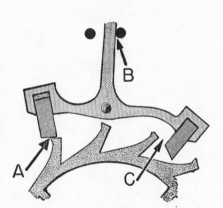
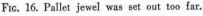
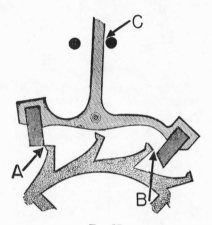

FIG. 16. Pallet jewel was set out too far.   FIG. 17.

surface a little before the pallet fork comes to rest against the banking pin. Figures 17, 18 and 19 show this in detail.

In Figure 17, the escape tooth A is just ready to drop off the entrance jewel. Escape tooth B is waiting to fall upon the exit jewel's locking surface. Notice that the fork has not yet touched the banking pin at C. In Figure 18, the escape tooth A has now dropped upon the locking surface of the exit jewel. Notice that the *lock B* is not deep at all. However, the pallet fork has not yet come to rest against the banking pin C as shown by the arrow.

In Figure 19, the pallet has slid down deeper into the escape tooth, making lock C now equal to specifications illustrated in Figure 14. The pallet is controlled and prevented from being drawn deeper into the escape tooth by the banking pin as shown by the arrow "A." If the banking pin were farther away, the fork would travel until it was stopped by the banking pin. This also would permit the pallet to be drawn deeper into the escape tooth with subsequent unwanted and excessive lock.

The action of the pallet as it slides down while the escape tooth presses upon its locking surface is called "Slide."

The amount of slide is controlled by the banking pins. Slide, in turn, is caused by the incidence of the angles of the escape tooth and the locking

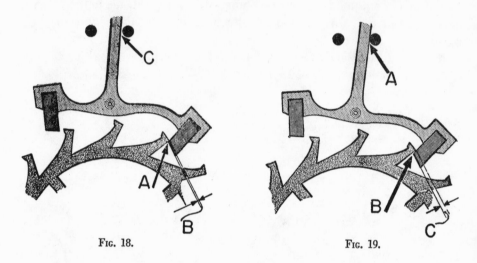

FIG. 18.   FIG. 19.

side of the pallet jewel. Pressure of the escape tooth against the locking surface of the pallet jewel here causes the pallet jewel to be "drawn" into the escape tooth. In escapement practice, this is called "draw."

When the jewel has been adjusted to meet the requirements of Figures 17, 18, and 19, excess shellac may be cleared away by cleaning with alcohol. The pallet is then ready for use.

SUMMARY

1. A pallet jewel should be replaced when it is chipped, broken, or missing.

2. Pallet jewels may be loosened by placing the pallet on a pallet warmer which melts the shellac.

3. The pallet is placed bottom side up on the warmer after the warmer is heated.

4. A new jewel should match the old one in thickness, length, and angle of lift.

5. Jewels are fitted to the pallet so that their lifting surfaces face the oncoming escape tooth.

6. "Lock" is the static form the escape tooth assumes as it rests against the locking surface of the pallet jewel.

7. Banking pins should not be moved unless the escapement is understood thoroughly.

8. "Slide" is the action of the pallet as it is drawn deeper into the escape wheel during "lock."

9. Slide is caused by "Draw." Draw is the result of the angles of the escape tooth pressing upon the pallet's locking surface.

10. Pallet jewels set too short will cause "tripping."

QUESTIONS

1. How is a pallet jewel removed?
2. How should a pallet jewel be inserted into the slot?
3. What is "lock"? How much lock is advisable?
4. What is a pallet warmer? How is it used?
5. In what direction should the lift surfaces of pallet jewels face?
6. What is the difference between an entrance jewel and an exit jewel?
7. How are jewels matched for angle, length, thickness?
8. How may a long jewel be shortened?
9. What is Draw, Slide?
10. What may prevent an escape tooth from dropping off the pallet jewel?
11. What is the result of too short a pallet jewel?
12. What may be the result from overheating a pallet warmer?
13. How much shellac is necessary to bind a jewel to its slot?
14. What is the rule about the banking pins?
15. What side of the pallet is up when it is on the pallet warmer?

XIII

REPLACING A ROLLER JEWEL

A roller jewel or jewel pin as it is sometimes called must be replaced when it is chipped, broken, or missing. (Figure 1.) A roller jewel should be replaced if there is too much play within the fork. (Figure 2.)

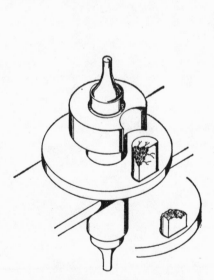

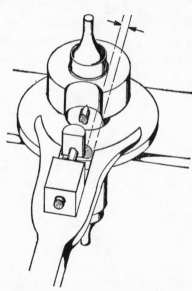

FIG. 1. Chipped or broken jewels such as these should be replaced.

FIG. 2. A jewel that permits too much play in the fork slot should be exchanged for a better fitting replacement.

The symptoms of a broken or missing roller jewel are simply that the watch will not run. Upon examination, the balance wheel will turn and vibrate when the watch is shaken without any movement of the pallet fork.

A chipped roller jewel pin should be replaced because the jagged edges of the chipped part will cut into the fork slot of the pallet, damage the fork and seriously impair the performance of the watch.

A jewel that is too thin permits excessive shake in the fork slot. Excessive sideshake dissipates the power of the pallet and the escapement. This prevents a good balance motion and makes position adjusting or good timing impossible.

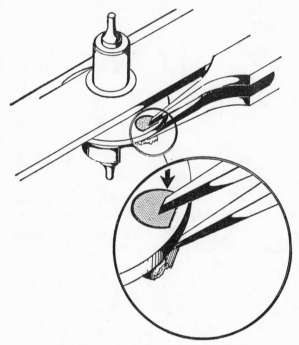

FIG. 3. How the old jewel may be ejected.

To remove a roller jewel for a replacement, the hairspring is removed from the balance although the roller table may remain upon the balance. Removing the rollers from the balance sometimes results in damage to the staff and endangers the balance truth and poise.

To remove the chipped or broken jewel in preparation for a replacement, a pair of strong-tipped tweezers is used. One point of the tweezer is braced against the underside edge of the roller disc and the other tweezer point is pressed down upon the jewel pin. This forces the broken jewel out in the direction of the pressure applied. The jewel should be ejected from the bottom side up toward the safety roller. (Figure 3.)

After the jewel is out, the hole is cleaned of all chips of jewel or shellac.

Selecting a New Jewel

There are generally three shapes or roller jewels used in modern watches:—D-shaped, triangular, and elliptical. The D-shape is by far the most commonly used. The shape is determined by the form of the hole in

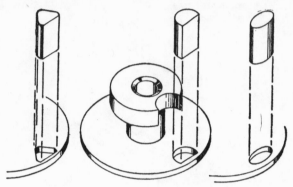

Fig. 4. Three main shapes of roller jewels.

the roller table or disc, as shown in Figure 4. However, the size or thickness is determined solely by the width of the pallet fork slot.

To choose a jewel pin merely because it fits the hole in the roller table snugly may result in the jewel binding in the fork slot or not being able

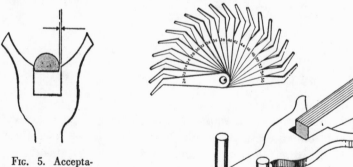

Fig. 5. Acceptable tolerance for a roller jewel.

Fig. 6. Roller jewel gauge and its application.

to enter at all. But too thin a jewel, as stated previously, wastes the power of the escapement.

A jewel should be selected which fits the fork slot with a minimum of sideshake. Therefore, the fork slot becomes the gauge for the jewel pin size. Figure 5 shows the satisfactory proportions of a jewel and the fork slot. Some watchmakers possess a set of "feeler gauges." (Figure 6.) These are a set of finger-like prongs graduated in size. Each prong is marked in hundredths of a millimeter. These are calibrated in graduations

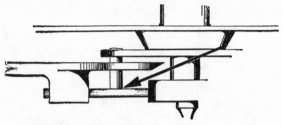

Fig. 7. Too long a jewel may butt the guard finger.

of .02 mm. The tip is inserted in the slot of the fork and the correct size noted by its fit.

The number on the gauge shows the proper thickness of the jewel. The advantage of this gauge is that the pallet need not be removed from the movement. It also eliminates the necessity of testing the jewel directly in

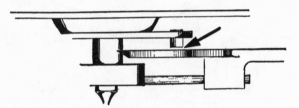

Fig. 8. Too short a jewel will not reach into the fork slot.

the fork. Often this results in the loss of the jewel due to its small size and difficulty in obtaining a firm grip upon it.

When a new jewel of the proper thickness and shape is selected, it is examined for length. A jewel that is too long will butt the guard finger as shown in Figure 7. One that is too short will not engage the fork. (Figure 8.)

If a jewel is too long, it may be shortened by grinding off the excess length with a diamond-charged lap as illustrated in the chapter dealing with the fitting of a pallet jewel. Some watchmakers place the over-long jewel on a hardwood block, across the grain, and with a sharp knife break off the excess length.

Grasping the jewel for insertion into the roller table hole is often a most delicate and trying experience for beginners, especially if the jewel is a tight fit in the shaped hole in the roller. (Figure 9.)

A simple method of inserting the jewel into the roller hole is suggested here. Place the roller jewel flat upon the bench plate. Wet the inside tip of a flat-tipped tweezer such as a Dumont No. 2. With the wet tip, touch the flat, polished end of the jewel. It will adhere to the tweezer as shown in Figure 10. The jewel may then be eased into its hole in the roller table. If the jewel is a snug fit, it may be squeezed into place by placing the opposite blade of the tweezer under the roller's edge and, by carefully pressing the tweezer blades together, the jewel will be pressed into its hole. The jewel is inserted from the safety roller side down toward the balance wheel. As in the fitting of the pallet jewel, the roller jewel is held in place by shredded or flaked shellac.

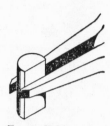

Fig. 9. Holding the roller jewel.

For the jewel to fit tight, the shellac when melted must flow into the smallest crevices of the hole. The shellac will be drawn into these crevices

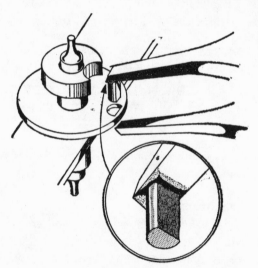

Fig. 10. A whetted tweezer tip will hold the jewel so that it may be "threaded" into its hole.

only if the roller table is heated first. Heat must be applied indirectly so that the balance is not burnt or overheated.

To do this, the roller table is held in a tool or clamp. The clamp must be made of a metal that will conduct heat from a wing of this tool projecting out of the clamp. This wing is held over the flame of an alcohol lamp. Shown in Figure 11 are a few commercial tools used for this purpose. Tool A will be used for the demonstration here. Figure 11C shows the tool used to hold the small roller jewel.

This tool has two small jaws which straddle the roller and clamp it by means of an internal spring. Each clamp has a V groove along its inner length to aid in keeping the roller disc in place.

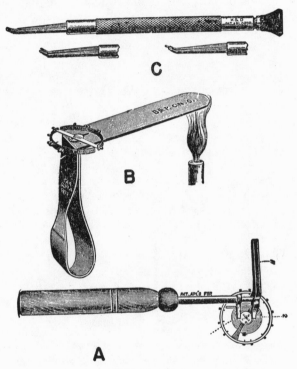

Fig. 11. Tools used in roller jewel setting.

To use this tool, the jaws are opened by pressing the bottom button. The disc or lower part of the roller table is then inserted between these jaws so that the balance arms are at right angles to the jaws of the tool and the roller table is in an accessible position. Releasing the pressure upon the button will permit the jaws to close upon the roller disc. Figure 12 shows this tool in action and method by which the wing is heated to melt shellac adjacent to jewel.

To hold the jewel in place, a small piece of shredded or flaked shellac is placed over the tip of the roller jewel as it emerges from the bottom of the roller. This amount must be small, about 1½ times the diameter of the roller jewel. (Figure 13.)

The wing of the tool is heated until the shellac melts into the crevices between the walls of the jewel hole and the jewel. The shellac must not melt into a ball. This indicates a need for more heat or else the surfaces are not clean. Furthermore, the shellac must melt over and into the jewel. Should the shellac melt and run into another spot such as the balance staff hub, it should be cleaned off and another piece of shellac placed over the jewel for another try.

Fig. 12. The flame from the alcohol lamp heats the wing which transfers the heat to the roller jewel and melts the shellac around it.

After this has been done satisfactorily, the balance is removed from the tool and inspected. A well-fitted roller jewel should be of sufficient height to engage the pallet fork. It left too short, it may not engage the fork slot, especially if the balance has more than the required end-shake. Here the balance in dropping upon the cock cap jewel may fall out of engagement with the fork slot as shown in Figure 8.

A well-fitted roller jewel should not incline in any direction. (Figures 14B and C.) It must remain upright as shown in Figure 14A. Should the roller jewel incline or sink into the roller during heating, it may be easily adjusted. A tweezer is heated over an alcohol lamp and applied to the cemented jewel. The heat from the tweezer transfers itself to the jewel which, in turn, softens the shellac holding it. The jewel may then be adjusted to its correct height and position.

Occasionally, a jewel of a certain size must be fitted into the roller disc, the hole of which is much larger than the jewel. The jewel, because it is much smaller than the hole, will not rest long enough to have the shellac melt around it. In such a case, the hole in the roller should first be filled with melted shellac. The roller jewel is then held in the heating tweezer over the alcohol lamp flame. The heated jewel will melt the shellac in the hole and upon cooling will re-main rigid and firm, held by the cooled shellac around it.

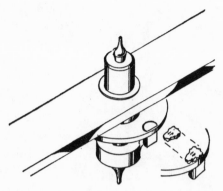

Fig. 13. Approximate size of chip shellac in proportion to roller jewel.

If the hole in the roller disc (table) is too small for the proper size jewel, it may be made large enough to accommodate the jewel. This is done by making a fine taper pin of iron or steel, filing one side off to fit into the roller hole. Some fine oilstone powder, mixed with oil, is then

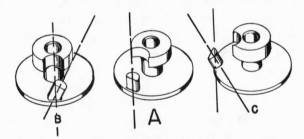

Fig. 14. The jewel must be set upright as in "A" and not incline either as in B or as in C.

applied to the D-shaped taper and inserted into the hole. Moving this taper in and out with the oilstone charged to it will soon enlarge the hole to the desired size and shape. (Figure 15.)

After the jewel is fitted and shellacked, it should be tested to see whether it is secure. This is done by holding the balance in one hand and grasping

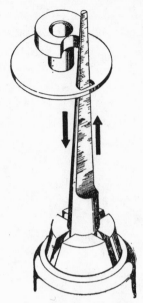

FIG. 15. Oilstone powder on a tapered lap may be used to enlarge a jewel hole.

the jewel in the points of the tweezer with the other hand. Moving the tweezer will soon reveal if there is resistance offered by the jewel or if it is loose. If it is loose, the shellac may be heated again in order to secure the jewel rigid to the roller table disc. After all excess shellac is chipped off, and the balance cleaned it is ready for placing into the movement.

SUMMARY

1. A roller jewel should fit the fork slot with a minimum of play.

2. Roller jewels should be replaced when chipped, broken or missing.

3. Roller jewels come in three shapes; oval, triangular and D-shaped.

4. Long jewels may be ground to size with a diamond-charged lap or skillfully cleaved.

5. The roller jewel may be held by whetting the inside flat tip of a tweezer or grasped with a fine tweezer.

6. A minimum of shellac should be used. Too much shellac will "run" and destroy the appearance of the job as well as the poise of the balance.

7. Too long a roller jewel will interfere with the pallet fork guard finger.

8. Too short a roller jewel may fail to engage the fork slot.

9. Jewels must be fitted upright, not slanted sideways, forward or backward.

QUESTIONS

1. Why should a chipped jewel be replaced?
2. Why should a roller jewel too thin for the fork be replaced?
3. How is a chipped or broken jewel removed?
4. What are the correct proportions of a roller jewel?
5. How is a new jewel inserted in the hole?
6. How much shellac should be used to fasten the jewel?
7. How is the shellac melted?
8. Why should a roller jewel be not too long; too short?
9. How may an inclined jewel be uprighted?
10. How is a jewel fitted to an over-size roller jewel hole?

XIV

CAUSES OF OVERBANKING

"Overbanking," or "out of action" as it is sometimes called, is the shifting of the pallet and fork from one banking pin to the opposite banking pin without having been engaged with the roller jewel.

When overbanking exists, the roller jewel on its return does not enter the fork slot. Instead it strikes the outside of the fork, as illustrated in Figure I. The force of the hairspring, in its effort to assume a position of rest, will keep the roller jewel against the outside of the fork horns, and the balance becomes locked in this position.

It is worth reviewing the normal action of the pallet at this time. After it has given impulse to the balance wheel the pallet comes to rest against one of the banking pins. The pallet is held against this pin during the motion of the balance wheel by the force of the escape wheel pressing against the locking side of the pallet jewel.

Should the watch receive a jar, the pallet may be jarred away from its position at the banking pin and might go over to the other banking pin except for the fact that the guard finger will hit the outer edge of the safety roller. After the influence of the jar has ceased, the pallet returns to its original position against the banking pin because of the pressure of the escape wheel tooth on the pallet jewel's locking side.

The *drawing* of the pallet back to its banking pin prevents the guard finger from remaining in contact with the safety roller and avoids any further interference with the normal oscillation of the balance. The pallet in a properly adjusted escapement can go from one banking pin to the other only during the time when the roller jewel enters the fork and the guard finger is in the "passing crescent" of the safety roller.

225

About the most common cause of overbanking is a short guard finger. In Figure 1, the pallet, shown in the overbanked position, is jammed against the banking pin opposite from the one on which it should rest. The pallet received a jar which caused it to travel along the path of the lower dotted arc. It was stopped by the banking pin at the left.

When the roller and balance returned in the direction indicated by the arrow, the roller jewel instead of entering the fork slot struck the outside of the fork and jammed the balance in that position.

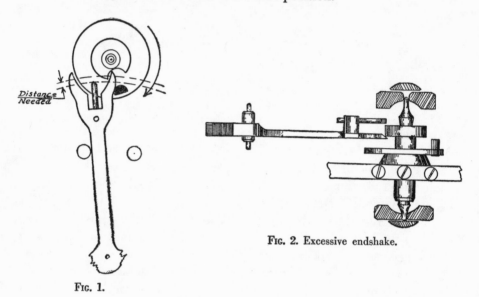

Fig. 2. Excessive endshake.

Fig. 1.

Had the guard finger been long enough, it would have hit the outside edge of the safety roller and prevented the fork from going over to the opposite banking pin. The extra length needed is shown by the upper arc.

Too Much Endshake

In Figure 2 the effects of too much endshake are shown. When the watch is in the dial-down position, the excessive dropping of the balance causes it to become disengaged from the fork. The safety roller now travels in a lower plane than the protecting guard finger. Any jar will permit the pallet to move over to the opposite banking pin and, as the balance returns, it will jam against the fork horns.

The correction in this case would be to reduce the endshake so that the balance is moved up and the rollers engage the fork and the guard finger. Some watchmakers accomplish this by bending the balance bridge. However, excessive bending may cause the regulator to interfere with the balance wheel.

Excessive Sideshake

Excessive sideshake may be the result of a cracked hole jewel or one that has a hole too large for the pivot; in turn, it causes other ill effects. One such fault introduced by this condition is overbanking. As shown in Figure 3, the pivot has shifted over into the cracked recess of the jewel. This has caused the safety roller to fall back of the arcs at which the safety roller and the guard finger normally intersect one another. A jar during the time the pivot is in the cracked portion of the jewel will cause over-

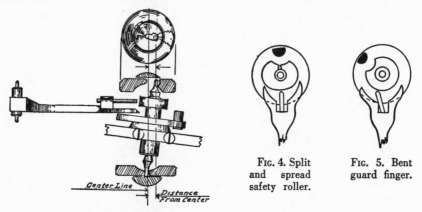

FIG. 4. Split and spread safety roller.

FIG. 5. Bent guard finger.

FIG. 3. Broken hole jewel.

banking. The jewel should be exchanged for one that fits closely around the pivot and is in good condition.

A Cracked Roller

A roller that is cracked may be due to forcing it on a staff with a very thick roller post. Such a roller has ceased to be perfectly round and is now oval. The safety roller shown in Figure 4 has become wider at the sides. The narrower sides offer a clear path for the guard finger should jarring occur when the guard finger is adjacent to this position.

Lengthening the guard finger will only cause it to rub against the wide parts of the ovalized roller. The remedy in this case is a new roller of the correct diameter and height.

Bent Guard Finger

A guard finger that is bent does not reach into the safety roller's orbit. As shown in Figure 5, this can cause overbanking. Furthermore, when a guard finger is only slightly bent, it will have too much freedom on one side and none on the other. This will cause the escape tooth to leave the locking surface of the pallet jewel upon the slightest jar and enter upon the lifting surface of the pallet jewel. The escape wheel will either push the pallet up and across to the opposite banking pin or the pallet may become jammed with the escape tooth on the pallet jewel's lifting surface.

Straightening the guard finger may overcome this error.

Distorted Roller Table

In Figure 6, the roller hole, being too large for the staff, was pinched or nicked with a triangular punch. As a result, the roller table was closed only on the bottom. The top shifted over to a side and a condition similar to Figures 3 and 4 resulted.

 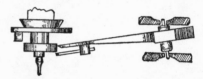

FIG. 6. Distorted roller (tipped).   FIG. 7. Pallet out of alignment.

The remedy is to replace this with a roller that fits the staff snugly and is of the correct diameter and height.

Excessive Side Shake in the Pallet Arbor Pivot

In Figure 7, the result of a pallet arbor that is too thin for the hole jewel, or a hole jewel that is too large for the arbor pivots, is shown. Here the pallet may move up and down. When it falls downward, the fork horns scrape the roller and the guard finger will touch the roller jewel. When the watch is turned over, the fork and guard finger will disengage from the roller and overbanking will result.

To remedy this situation, pallet jewel holes with close-fitting tolerances together with correctly shaped pallet arbor pivots must be fitted. Jewels for the pallet arbor pivots must fit and be set plumb over one another; otherwise, the pallet may not lie at right angles to the axis of the balance staff.

Crushed Rollers

A roller table that was crushed by driving it too hard on a roller post will cause the safety roller to encroach on the level occupied by the roller jewel pin as shown in Figure 8. The safety roller is now below the guard finger and any jar will permit free movement to the pallet, aided by the escape wheel. Overbanking results.

When a crushed roller is the cause of overbanking, it must be replaced with a new one which meets all the requirements of height, diameter, and fit.

Bent Pivots

When the pivots are bent as shown in Figure 9, the line of centers or staff axis has shifted from a point where the guard finger does its protective work to a position beyond its influence.

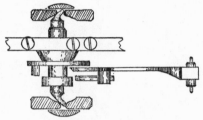

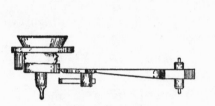

FIG. 8. Crushed roller.   FIG. 9. Bent pivots.

This illustration shows the balance and rollers moved away from the pallet fork and the guard finger. As the balance turns (if it could), it will scrape the guard finger because the rollers are now turning on an eccentric staff. Then, again, as they continue their motion, the rollers will turn away from the pallet fork and guard finger. In this position, the watch may overbank. Although this illustration shows a rather exaggerated condition, even a slightly bent pivot may cause overbanking.

If the pivot could be straightened, it would prevent another occurrence of overbanking. If the pivot cannot be corrected satisfactorily, a new staff must be inserted.

The Correct Guard Finger Length

The correct length of a guard finger is shown in Figure 10. The arc described by the guard finger passes through the small, circular path of the safety roller. Also, when the guard finger is of the proper length, the lock A will always be greater than the space B.

Testing for overbanking is done while the pallet is against the banking pin and the watch is wound up a trifle. The balance is twisted around until the roller jewel is out of the fork slot similar to the position shown in Figure 10. While the balance is held in this place by the soft tip of the index finger, the point of a fine tweezer or needle is placed at the pallet

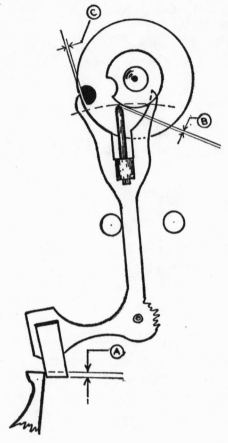

FIG. 10. Proper guard finger length.

jewel and the pallet is pushed by the pointed tool upward in an effort to move it away from the banking pin.

If the distance B is greater than space A and the roller jewel is beyond the influence of the fork horns, the escape tooth will enter upon the lifting surface of the pallet jewel and push the pallet over toward the other banking pin. If this occurs, the guard finger must be lengthened. This test is applied when the fork rests against either of the banking pins and should be applied to both positions.

A guard finger of the proper length shows up in this test by permitting the pallet to be moved up a bit without lock (distance A) disappearing. During this manipulation the guard finger will touch the safety roller, and further upward movement of the pallet jewel will be prevented. Distance B, of course, disappears when the guard finger touches the safety roller.

This test should be applied in as many positions as possible so that all the causes of overbanking may be discovered.

SUMMARY

1. Overbanking takes place when the pallet moves over to the opposite banking pin without having had any contact with the roller jewel.

2. The most common cause of overbanking is a guard finger that is too short.

3. Short guard fingers may be pushed out farther or may be stretched.

4. When a watch is overbanked, the roller jewel rests outside the pallet fork and against one of the fork horns.

5. A bent pivot will cause the safety roller to swing away from the path of the protecting guard finger.

6. A cracked jewel may also be the cause of overbanking.

7. A test for overbanking is to try to unlock the pallet jewel while the balance and roller jewel are moved outside the fork slot.

8. A bent guard finger may cause overbanking by permitting excessive and uneven guard finger freedom.

9. The pallet should be tested in every position to discover all causes of overbanking.

10. The guard finger must be long enough to intersect the roller diameter yet permit the pallet fork to pass during the time when the roller jewel is in the fork slot.

QUESTIONS

1. What is overbanking?
2. Name eight causes of overbanking.
3. How is the watch tested for overbanking?
4. What is the actual cause that permits the pallet to travel to the opposite banking pin?
5. What causes the roller jewel to stay jammed against the outside of the fork horns?
6. How may a bent pivot cause overbanking?
7. What is the relationship between lock and guard finger freedom?
8. What change takes place in the shape of a roller when it is cracked?
9. If the safety roller is forced lower into the plane of the roller jewel, what may take place?
10. What is the proper length of the guard finger? How is this tested?

XV

ADJUSTING PALLET-GUARD FINGERS

The guard finger is the spear-shaped projection protruding from the pallet fork or lever. The purpose of the guard finger is to provide a safety action of the lever during the split moments when the watch is jarred. This is shown in Figures 1, 2, 3 and 4. The lever in the case of sudden shock tends to move away from its position of rest against the banking pins and move towards the opposite side.

Figure 1 shows how the guard finger appears through the eye loupe of the watchmaker. The lever is resting against the banking pin B. In this case there is sufficient freedom or space between the edge of the guard finger and the edge of the safety roller C. The space A is much less than the lock B. This is correct so that when the watch is jarred and the lever moves away from the banking pin (Figures 3 and 4), the guard finger does come in contact with the edge of the safety roller. In doing so, the pallet moves up and reduces the lock now reduced to the proportions shown in E. This reduction in lock is directly proportioned to the loss of space C and increase in space D.

In these first four illustrations it is shown how a guard finger of the correct length will provide safety action and prevent overbanking. The only way the lever can go to the opposite banking pin is during the time when the roller jewel enters the fork slot and the guard finger enters the safety roller crescent or "passing hollow" in its orderly and normal processing from side to side.

Figures 5 and 6 show the same pallet with the guard finger shortened to explain the lack of safety action. The guard finger is so shortened that, when the lever is jarred, the shortened guard finger describes an arc (F) shown in the dotted lines, which does not intersect the circumference of the safety roller. The pallet jewel has moved up to such an extent that the escape wheel tooth has entered upon the lifting surface of the pallet jewels and forced the lever over to the opposite banking pin with the opposite

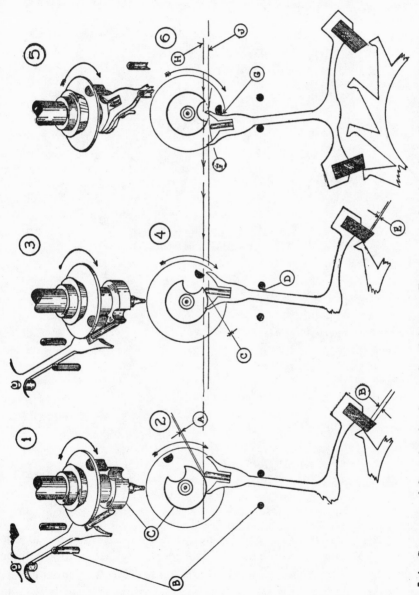

Left: Correct guard finger length. *Center:* Guard finger in Fig. 1 now touches the safety roller due to a shock, but lock E remains. *Right:* What would happen to pallet in Fig. 2 if guard finger F were too short. It is now "overbanked."

pallet jewel now locked against another escape wheel tooth. The balance, in the meantime, has been moving in the direction shown by the arrows concentric with the roller table and has approached the pallet fork slot.

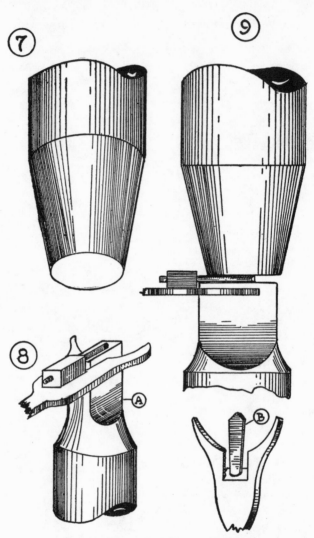

Peening or stretching the guard finger.

The roller jewel cannot enter the fork slot as it should because the fork is resting upon the opposite banking pin. The roller jewel, therefore, bumps violently against the outside of the fork and comes to rest against it (G), remaining there because the hairspring is still pulling it towards its line of "rest" or neutral position. This condition is called "overbanking." The watch, of course, has stopped. The dotted line H represents the true or correct length of the guard finger; in this particular case, J is the line representing the length of the short guard finger pin compared to H.

Stretching a Guard Finger

To correct a guard finger that is just a bit short, the guard finger may be stretched by resting the guard finger upon a miniature anvil (Figure 8A) which is inserted through the fork slot and allowing the guard finger to rest upon it. This little anvil may be ground or filed to shape from old phonograph needles or punches, and hardened. If the anvil fits the fork slot just right without too much play or without being too tight, the job will be done better. The finished anvil or stump may be held in a hole punch reversed in the staking set to form the "stump." A flat punch (7) is brought down over it as in Figure 9, while the pallet is held between the index and thumb of the hand. The flat-faced punch is brought down over the guard finger and tapped lightly with the hammer. The effect (B) will be to flatten the finger but lengthen it as well. The guard finger in this case, when not flattened excessively, is a desired effect as well, since it presents a thinner surface to the safety roller with resultant loss of unwanted friction and less chance of bending sideways, which is the direction of the strain upon all guard fingers of double rollers.

Removing the Guard Finger

There are many styles of attaching guard fingers to pallet forks. Except in the most expensive watches, they take on the forms as shown in these drawings. When a guard finger becomes broken, mutilated, or too short for stretching, it should be replaced with a new tapered pin which will become the guard finger. Removing the guard finger often presents a problem both to apprentices and journeymen. Figures 10 and 11 show two methods of removing the small pin from the "boss" of the pallet fork. Figure 10 shows a strong pair of tweezers braced one end over the end of the guard finger (at times broken off to this length to facilitate removal) and one blade of the tweezers against the bottom of the boss. The tweezer, when squeezed, will eject the pin or loosen it. Should it appear to become tighter, reverse the process as the original taper pin may have been inserted from the fork end towards the pallet arbor. Figure 11 shows how the guard finger is pushed out when the pin is flush with the opening in the "boss" or box. The fork is rested against the stump from the staking set on a bench block; a strong, sharpened needle set into a piece of pegwood will push out the stubborn pin. Sometimes both methods described in Figures 10 and 11 may have to be used.

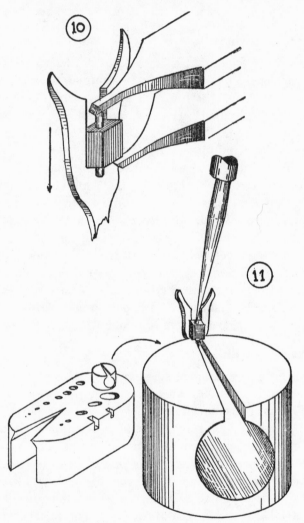

Methods of ejecting a broken guard finger.

A new guard finger is made from very small brass or nickel alloy taper pins. These may be purchased from your material jobber and are also used as replacement regulator pins, stud pins and for numerous other purposes.

The tapered pin is pushed through the hole in the fork boss from the pallet arbor side out toward the fork slot (Figure 12). This should be pushed in as far as it will go with pressure from the tweezers. A little nick is cut near the fork boss with a screwhead file so that it may be broken

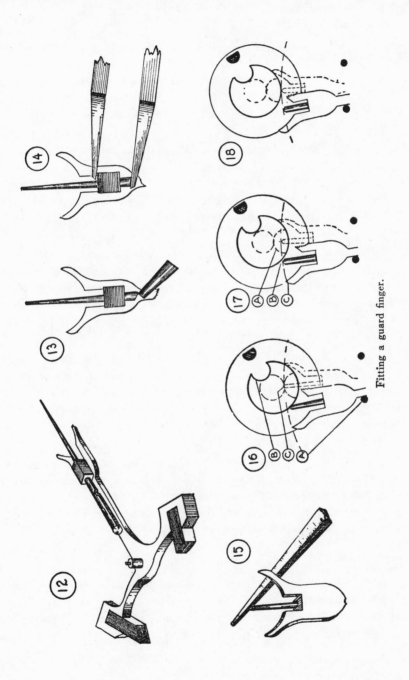

Fitting a guard finger.

off conveniently as in Figure 13. It is made secure by bracing the strong point tweezers as in Figure 14, and squeezing it in further. It is always best after jamming the pin tight to leave some of the pin sticking out of the fork boss (just a little less than is shown in Figure 14).

When filing the guard pin shorter to size, a square needle file with a safe edge is used. This file is placed flat on the fork horns so that the safe edge touches the fork horns. In this way the filing operation will not harm the fork. The proper angle to file the guard finger is shown in Figure 15 with the filing edge against the guard finger and the safe edge of the square needle file resting on the pallet fork horns. When the file is applied to the other side of the guard finger, the resultant angle will be close to 90° or one corner of the guard finger will be at right angles to the other.

Proper Length of Guard Finger

By studying illustrations from 1 to 6 and 16, 17 and 18, an idea of the proper length of a guard finger may be had. Figure 16 shows the guard finger has been left too long and is rubbing against the safety roller. This will not permit the balance to rotate freely. The banking pin (A) should never be tampered with except by watchmakers experienced in escapement adjusting. Should the banking pin be farther to the left, the guard finger may move away from the edge of the safety roller, but, when the fork and guard finger are entering the passing hollow B or safety roller crescent, the tip of the guard finger may butt against the innermost corner of the crescent C as shown in the phantom views of the dotted lines. Figure 17 shows the guard finger with an effective length. The lever is resting against the banking pin and the guard finger has sufficient clearance (B) from the safety roller (see Figures 2 and 4). The guard pin enters the crescent A, and the path of the guard finger C intersects the edge of the safety roller. Figure 18 shows the guard finger filed again too short. The guard finger is too far away from the edge of the safety roller, leaving the pallets and escape wheel no margin of lock safety. Also the path of the guard finger shown in the dotted line barely intersects the edge of the safety roller, leaving no bar to the unorderly procession of the guard finger from one banking pin to the other without benefit of action with the balance roller jewel.

Single Rollers

Single rollers are fast disappearing from the market, as today the double-roller escapement prevails. However, there is still a considerable number of these watches yet around to merit some mention. The safety action of this roller is in an upright pin projecting at right angles from the pallet fork. This acts against the edge of the roller which, in this case, serves both as a body for the jewel pin and the safety roller. Figure 19 explains the action. Figure 20 shows how the roller moves across the path of the fork. The guard pin enters the crescent, and the roller jewel receives an impulse from the pallet. Figure 21 shows the top view of this same action. Figure

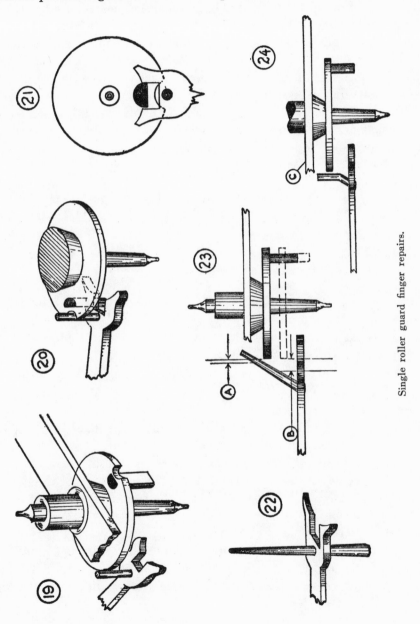

Single roller guard finger repairs.

22 shows how the new taper pin is inserted. The pin is inserted and broken off near the edge of the underside of the pallet fork. It is then driven farther in and tightened by placing the fork into the staking set with the thin part of the tapered pin in a hole of the staking die and driving the pin down farther with a small flat-faced punch. The pin is then shortened to size. These pins should be long enough to engage the roller edge and yet not too long that they catch the balances arms (C, Figure 24).

Bending the guard finger simply forward to overcome overbanking will not do the job since the distance between the edge of the upright guard finger and the edge of the roller varies when the endshake increases. Figure 23 shows this. The space (A) is the distance between the guard finger when the watch is in the dial-up position. When the watch is upside down, the endshake brings the roller table lower, and the distance then becomes as great as shown in space B. The proper bending practice is shown in Figure 24.

Summary

1. The guard finger is a projection from the pallet fork designed to offer a safety action of the pallet.

2. In the event of sudden shock the lever moves from its position of rest against one banking pin to the opposite side if the guard finger is not properly adjusted.

3. If the guard finger is of the proper length, it will come in contact with the edge of the safety roller during the moment of a shock. This prevents overbanking.

4. If the guard finger is too short, the safety roller has no contact with it. The roller jewel does not enter the fork but comes to rest outside and against it because of the pull of the hairspring. The watch has now stopped. This is called "overbanking."

5. A guard finger may be stretched longer by fitting it over a small anvil and tapping it lightly with a flat punch.

6. The guard finger may be removed either by use of tweezers or of a sharpened needle set into pegwood or both.

7. The replaced guard finger may be made from small brass or nickel-alloy tapered pins.

8. The inserted tapered pin should be pushed toward the fork slot, jammed in, and then filed with a square needle file to correct length.

9. With the single-roller escapement the guard finger may be replaced by a tapered pin inserted from the underside of the pallet fork. It is then shortened to correct length.

QUESTIONS

1. What is meant by "overbanking"?
2. How does a guard finger prevent "overbanking"?
3. Explain the consequences of too short a guard finger.
4. Why does a watch stop when "overbanking" occurs?
5. How would you lengthen a guard finger?
6. What is the condition of the guard finger after lengthening? Is this desirable? Explain the reason for your answer.
7. Describe two methods for removing a guard finger.
8. Name two metals which are satisfactory for use as guard fingers.
9. Explain how you would replace the guard finger. For filing, why is a file with a safe edge recommended? What precautions must be taken with the angle of filing?
10. What happens when the guard finger is too long? When filed too short?
11. How would you replace a guard finger for use with a single-roller escapement?
12. Why can't "overbanking" be remedied by merely inclining the guard finger?

XVI

HOW TO PUT A WATCH IN BEAT

A watch is "in beat" when a straight line may be drawn through the balance staff, roller jewel and pallet arbor during the time when the watch is at rest and the power in the mainspring is entirely released.

A watch in beat at the time when all power is released should have its escape tooth flush upon the lifting surface of the pallet jewel as shown in Figure 2, point A. Such a watch, in otherwise good condition, should start off by itself with a slight turn of the crown. This is further explained by the fact that in this neutral position the mainspring will motivate the escape wheel.

Since the lifting surface of the escape tooth is flush upon the lifting surface of the pallet jewel, a slight winding of the mainspring should cause the pallet to be lifted by the started escape wheel. This, in turn, will act upon the balance roller jewel which is in the center of the fork slot, giving an impulse to the balance.

A watch out of beat has an uneven tick. The balance will swing in uneven arcs, favoring one side of the swing and weak on the other. Such a watch may stop when the mainspring is almost run down. Furthermore, the watch will not start by itself but may have to be shaken or started manually. The roller jewel pin comes to rest out of the fork slot, and the pallet rests near one of the banking pins.

To test the watch for beat, release all the power in the mainspring and observe the position of the roller jewel as the balance is permitted to come to rest. If the watch is out of beat, the balance wheel, governed at this time by the hairspring, will come to rest with the roller jewel out of line with the slot in the pallet fork. The fork in such an event may not rest midway between the banking pins but rather nearer to one side or touching a banking pin.

Figure I shows an escapement out of beat. Here the balance has come to rest with the roller pin A off the line of center and almost out of the

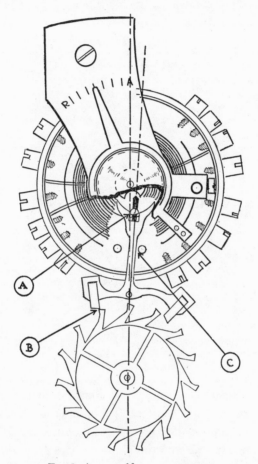

FIG. 1. An out-of-beat escapement.

pallet fork slot. The pallet is close to the banking pin C, and the escape tooth rests against the let-off corner of the pallet jewel B. This watch will not start off by itself when wound because the tooth will fall off the jewel and will not give the balance any appreciable push to return it with sufficient power and momentum to unlock the escape tooth. Thus, the escape tooth will remain locked on the exit pallet jewel.

The watch illustrated is only slightly off beat and does not indicate an exaggerated example.

How to Find the Position of Beat

To obtain the correct beat position, release most of the power in the mainspring, leaving only enough to move the escape wheel weakly. The balance will then be governed by the hairspring, coming to rest at a posi-

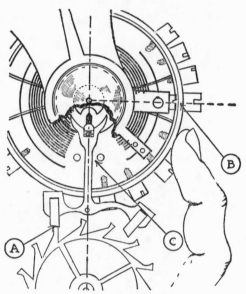

Fig. 2. The index finger moving the balance to the "in beat" position.

tion where the hairspring is neither wound nor counterwound. Then place the index finger against the edge of the balance as shown in Figure 2. Turn the balance until the roller jewel is in the middle of the fork slot and the pallet is equidistant between the banking pins as shown at C.

Since there is still sufficient power in the watch to move the escape wheel, the wheel will now be flush upon the center of the lifting surface of pallet jewel A.

For the next step, the eye must draw an imaginary line from the balance staff center outward through the middle of the stud and across the balance wheel rim. In Figure 2, the line B is pictured as crossing the balance in the middle of the balance arm, although this is not the position in all watches. This imaginary line may be placed elsewhere, depending, of course, upon the shape of the balance bridge and the relative position of the hairspring stud.

Note the position and exact point on the balance where this line intersects. It is here that the hairspring stud and balance must come to rest when the watch is unwound. If the hairspring collet is twisted on the staff so that the stud will come to rest adjacent to point B, the balance will be in beat.

In Figure 2, the balance has been nudged to the beat point. After noting this position, the balance is released and it will then return to its position of rest. The original discrepancy can be determined now.

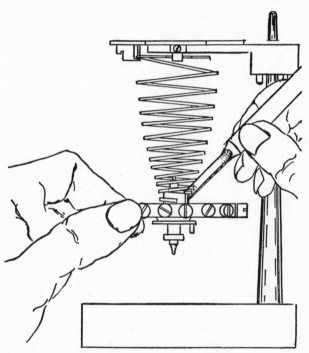

FIG. 3. A balance tack or scaffold used to hold the balance bridge while the collet is twisted.

An aid in noting the beat point is to make a strong mental picture of the exact position of beat. For instance, one should note that the line crosses the stud and across the balance in the middle of the third screw above the balance arm; or, again, this line might cross midway between two certain screws, favoring one or the other.

Some watchmakers mark the balance at the "beat point." Some manufacturers also follow this practice but it should be avoided if the marking will mar the appearance of the balance wheel.

Figure 3 shows a method of twisting the collet while the balance is attached to the bridge. This tool consists of a tapered steel pin about 1" or more in height fastened at the bottom to a brass block which measures 1" square and $\frac{1}{4}$" thick. The pin sticks up through the screw hole in the balance bridge, and the balance, thus hanging, permits the use of both hands for the work.

One hand holds the balance. In the other hand is held a needle set into a piece of pegwood. This tool is pictured in Figure 2, Chapter 7. The needle is ground so that its tip becomes a miniature jewel screwdriver.

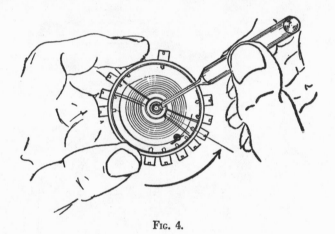

FIG. 4.

The tip is then inserted into the slot in the collet which is twisted so that the stud on the other end of the hairspring is shifted to the position previously noted as the "beat point."

Figure 4 shows how the hairspring may be shifted to the beat position when the balance is disengaged from the bridge.

Points to Be Checked

Figure 5 shows an escapement in beat. Check all the points leading to the circled letters at the right. B is the staff center. C is the roller (jewel) pin. D is the pallet. B, C, and D are situated on G, the line of centers. E shows the escape tooth flush against the pallet jewel's lifting surface. (If the escape tooth were in similar conjunction with the exit jewel, it would also be in beat.) F is the escape center.

H shows that the pallet fork is midway between the banking pins. If a pallet is midway between the pins during rest it is a further indication of being in beat. Some watchmakers use this as a criterion to guide them in setting the escapement in beat. While it is a worthwhile aid, this should

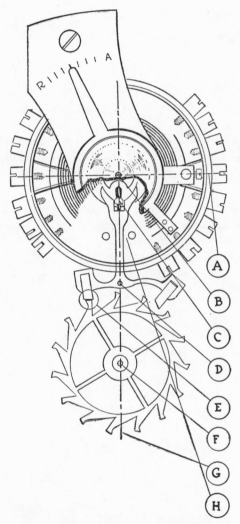

Fig. 5. Points for "in beat" checking.

not be the sole guide for the operation. If we observe this illustration we can readily see that, whereas B, C, and D may be on line G, one of the banking pins could easily be shifted or bent farther to one side. This new position would then mislead a watchmaker dependent upon this one clue.

In such a case after following all the directions noted above, the escapement should be checked first and the banking pins realigned later.

QUESTIONS

1. What is meant by "out of beat"?
2. When is a watch in beat?
3. Why should a watch be in beat?
4. How is the beat point found in a watch?
5. How is it possible to put a watch in beat without removing the balance from the bridge?
6. Under what conditions will a watch stop when it is out of beat?
7. How is the balance bridge held while the hairspring is being shifted?
8. Describe the tools used to shift the hairspring collet so that the stud assumes the beat position.
9. Name all the checking points to be observed in putting the watch in beat.
10. If all checking points are found correct but the pallet does not rest between the banking pins, what procedure should be followed?

XVII

COLLETING AND STUDDING A HAIRSPRING

Part I: Pinning the Hairspring to the Collet

There are occasions when it is necessary to "collet the hairspring." By this is meant the fastening of the center of the spiraled hairspring to a brass split ring called the "collet." The attachment of the hairspring is made possible by inserting the end of the innermost coil into a small hole in the rim of the collet. It is made fast by means of a tapered brass pin wedged into the hole.

The operation of pinning the hairspring to the collet, or "colleting" as it is called, is resorted to when it is necessary to change a hairspring to fit the balance or to replace a broken or cracked collet. It is also utilized when it is needed to change the position of the "pinning point."

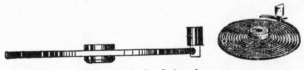

Fig. 1. A flat hairspring.

This chapter deals only with the immediate needs for fastening the hairspring to the collet rather than other considerations such as truing, centering, vibrating, or leveling the hairspring.

There are two types of hairsprings used in the ordinary modern watch—the flat and the Breguet (or overcoil).

Figure 1 shows the "flat" hairspring, one whose spiraled coils are on a flat, level plane. The other, Figure 2, shows the Breguet, so called after

249

its inventor. This spring starts out similarly as the flat hairspring does, but, just at the beginning of the last coil, it rises and turns in and above the main body of the spring. The last raised coil is most often smaller in diameter than the lower coil.

FIG. 2. Two views of an overcoil or Breguet hairspring.

The collets for these hairsprings have their individual characteristics necessary to their proper functioning. The collet of the flat hairspring, as shown in Figure 3, has the pinning hole drilled along its rim nearer the upper part of the collet.

FIG. 3. Collet for flat hairspring.

FIG. 4. Collet for Breguet hairspring.

The collet for the Breguet hairspring is pictured in Figure 4. Notice that this collet has its pinning hole drilled along the rim closer to the bottom. This is done to provide more room between the balance wheel and the balance bridge for the two levels of hairspring. The collets have beveled holes to facilitate their placement on the staff.

The tools required for colleting the hairspring are: collet holding devices, a No. 5 tweezers as well as a sturdy, stubby pair (Figure 14), a pointed steel needle, and a screwhead file.

Figures 5 and 6 show two simple tools which will hold the collet firmly during the manipulative operations. They permit clear vision and easy handling. Figure 5 represents a round, tapered steel pin filed so that the back remains round while the sides are triangular-shaped. The collet is placed over this tapered pin so that the split side fits over the sharp edge of the triangular section. This will make it secure and prevent its turning under pressure against its edge (see Figure 7).

Figure 6 indicates a more elaborate tool, the better of the two. A is a brass rod about 5 mm thick, drilled and tapped to receive the bevel-headed screw which holds down the collet. The base upon which the collet rests is countersunk to prevent the collet from spreading because of pressure of the bevel screw.

In order to obtain a clear background so that the coils of the hairspring may be plainly seen, a seconds' bit of an old porcelain dial is placed on a shoulder of this tool as shown in B. This seconds' dial may be shellacked to the shoulder of A so that it may be kept in place.

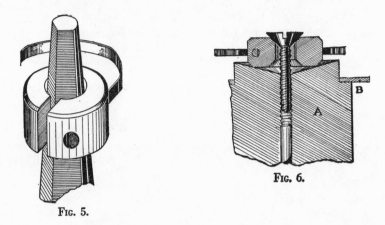

FIG. 5.

FIG. 6.

In removing the hairspring from the collet, the old taper pin must be removed. This is wedged into the hole in the collet together with the hairspring. To remove this pin, a sharp, strong needle is placed against it and

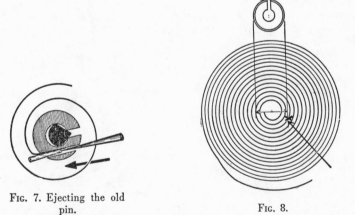

FIG. 7. Ejecting the old pin.

FIG. 8.

pushed outward. As shown in Figure 7, the needle ejects the pin from the inside of the coil outward toward the main body of the hairspring. This is done since all tapered pins are originally set into the collet holes from the hairspring side. When this old pin is removed, the hairspring will come out easily.

If a new hairspring is to be used, it must be prepared so that the innermost coil will fit correctly into the collet: Figure 8 shows a new hairspring with the inner coils crowded so that it would be impossible to fit this spring into the collet. Therefore, some of the inner part of the new spring may have to be broken off to permit the spring to be fastened to the collet.

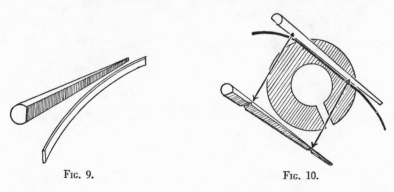

FIG. 9.   FIG. 10.

To determine how much of this inner coil should be removed, place the collet over the hairspring as in Figure 8. The diameter of the collet will touch two coils of the new spring. Where the collet's diameter touches the outermost of these two coils (indicated by X in Figure 8), is the spot where the spring should be broken.

Brass taper pins for colleting and studding may be made or purchased. These are generally about $\frac{1}{4}''$ long. The hairspring is secured in the hole

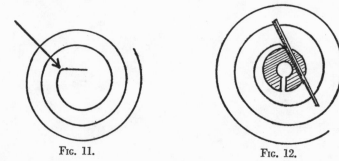

FIG. 11.   FIG. 12.

of the collet by wedging this pin against the hairspring. In order to obtain a good fit against the hairspring, one side of the pin is filed slightly flat as in Figure 9. (The old pin should not be used.)

This pin is then inserted into the collet with a piece of trial hairspring as shown in Figure 10. The pin should be pushed in as far as it will go without jamming it tight. In this position it is nicked with a screwhead file at the points indicated by the arrows. This will facilitate subsequent breaking off of the pin. The grooves are so placed on the pin that, when it is broken off in the latter stages of the colleting operation, both ends will be flush with the holes in the collet after being pushed in.

The pin and the sample piece of the hairspring are then removed from the collet.

The inner coil of the hairspring must be bent inward so that it will enter the hole in the collet without distortion and facilitate the centering of the hairspring around the collet. The bend should not be too sharp or use too much hairspring. This end must be straight so that it fits into the hole in the collet wall easily. Such a bend is shown in Figure 11.

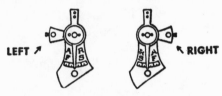

FIG. 13.

The hairspring is inserted into the collet hole so that it turns outward away from the split in the collet. Care should be exercised that the collet is right side up, taking into consideration the specifications shown in Figures 3 and 4. Furthermore, the hairspring must be inserted so that it spirals clockwise if it is a "right" hairspring or counterclockwise if it is

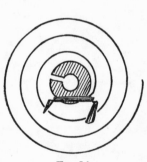

FIG. 14.

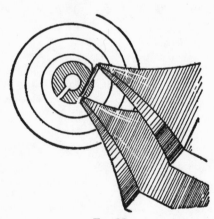

FIG. 15.

a "left" hairspring, depending upon where the stud is placed on the balance bridge (see Figure 13).

When all the aforementioned has been checked, the nicked taper pin is pushed in alongside the inserted hairspring so that the filed nicks face inward and against the inserted part of the spring. While the pin is eased into this hole, the hairspring is adjusted so that it will remain level as the taper pin is pushed in tight. This is shown in Figure 12.

When the pin has been pushed in as far as it will go, the excess length of pin is broken off in a direction away from the collet as shown in Figure 14. To make the pin secure, a short, pointed pair of tweezers is used. One of its points is braced against the side of the collet hole. The other is placed over the thicker end of the taper pin which may be protruding. Squeezing the tweezers will push the taper pin in tighter. If the pin was properly prepared, it will be flush at both ends, no further "clipping" being necessary. This is shown in Figure 15.

Colleting Bonded (Elgin-Type) Hairsprings

Hairsprings as in modern Elgin watches are fastened to their collets and studs by a resin. Should these need colleting, remove all of the bond material from the spring and collet slot. Clean both slot and spring in alcohol to provide a clean surface. Next, press the collet into the surface of a pithwood block so that the hairspring when placed in the slot, will lie level, resting both in slot and on pithwood. Place a shred of shellac on the juncture. Next, place the heated tang of a pallet or roller warmer over the shellac and spring, causing it to melt and bond these together. If the slot is inadequately filled with shellac, repeat the process shown in Figure 16.

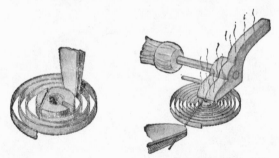

Elgin National Watch Co.

FIG. 16.

QUESTIONS

1. When is it necessary to collet a hairspring?
2. Describe briefly two common collet holding devices.
3. Explain how the tapered pin should be prepared for pinning the hairspring.
4. How can you determine how much of the inner coil of the hairspring must be broken off? Why is this necessary?

5. Explain briefly how the hairspring is pinned to the collet.

6. How is the excess pin removed from the collet hole?

7. How is the hairspring made level with the collet?

8. Make a diagram showing the coils of the hairspring in relation to the collet.

9. How is the hairspring centered in the collet?

10. Explain fully the kind of collet you would use for these hairsprings.

Part II: Studding a Hairspring

"Studding a hairspring" is an operation in which the stud is connected and secured to the outer terminal of the hairspring.

A hairspring may have to be studded after it has been vibrated to the balance wheel and has already been colleted. This operation may also be required when the hairspring has been retimed following a break in the hairspring close to the stud.

Types of Studs

There are about six types of studs in general use. These are shown in the following six illustrations.

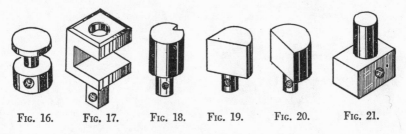

Fig. 16.   Fig. 17.   Fig. 18.   Fig. 19.   Fig. 20.   Fig. 21.

Figure 16 shows the mushroom or collar-button stud used on higher grade watches. It is also called a "floating" stud. This stud fits into a fork-like pier in the balance bridge.

Figure 17 shows the box type stud, also used on good quality watches. A screw threaded into the balance bridge holds it in place. This is also a floating stud. Floating studs are those that are attached to the bridge after the balance is in the watch.

Figure 18 shows the cylindrical or plain round type. This is the most commonly used stud and is found in a very great majority of Swiss watches. The V slot accommodates the stud screw and keeps the stud from twisting in the hole.

Figure 19 shows a triangular stud, used mainly with overcoiled hairsprings. This type is used in some American watches. The triangular shape, fitted closely into a hole in the balance bridge, keeps this stud firmly in place.

Figure 20 shows the D-shaped stud, a variation of the triangular shape. This is used in many of the better Swiss watches.

Figure 21 shows the friction stud. It is usually made of brass whereas the others may be made of steel. This stud is used mostly in cheap pin pallet watches and with cylinder escapement watches. This stud is merely pushed friction-tight into a close-fitting hole in the bridge. There is no set screw to insure the stud from slipping in its hole.

Holding the Stud

Holding the stud so that the hairspring may be threaded into its hole may be done by many methods. It is important, however, in order to permit accurate studding and to prevent damage to the spring from slipping and mangling, that the stud be held so that a firm grip is maintained, requiring but one hand and leaving the other free for manipulation.

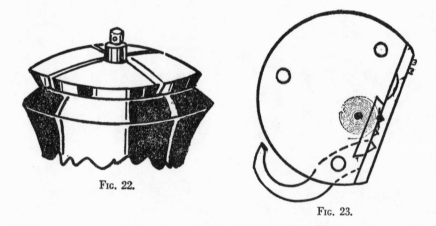

Fig. 22.

Fig. 23.

Some watchmakers prefer to use the balance bridge for this purpose. However, sometimes the balance bridge is so small and odd-shaped that it does not permit the fingers to grip the stud firmly and thus it hinders facile manipulation.

A pinvise, holding the stud as in Figure 22, permits a firm grip of the stud, clear vision, and complete freedom of one hand to manipulate any auxiliary tool.

For specialized work in studding, a studding table, shown in Figure 23, is unsurpassed. It is in a form of miniature vise. This table has a steel piece dovetailed into the front edge of the table. There are three or four small V grooves graduated in size in the edge of this steel piece. A thick steel spring is held against the dovetailed piece. A lever, pivoted in the corner of the studding table, pushes the spring-like clamp away from the studding table. This releases the stud or permits one to be fastened by the clamping vise-like action of the strong spring.

The table is about ¾″ wide and is supported by three legs. The advantage of using the table is that the top of the stud is held by the table with its cross hole clear. The hairspring may rest upon the table as shown. The spring is then easily threaded through the stud hole and easily fastened.

Removing a Hairspring from the Stud

Very often when a hairspring must be studded, the stud still contains part of the old spring and a short taper pin.

Fig. 23A. A hairspring studding table.

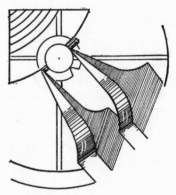

Fig. 24.

To remove the hairspring from the stud, place the stud into the pinvise as shown in Figure 24. Place one of the points of a short, stubby-pointed tweezer over the thin end of the slightly protruding taper pin. The other point of the tweezer is placed beside the thicker end of the pin and braced against the stud. By squeezing the tweezers, the taper pin will become loose. However, often this is not enough to eject the pin or to loosen it sufficiently so that the pin may be removed with the tweezers. If this is the case, take a fine-pointed steel needle or pin and push out the taper pin as shown in Figure 25.

Sometimes it is difficult to ascertain which is the thinner end of the taper pin. This is especially true when the pin has been cut or broken off flush at both ends. To eject such a pin, push it out from the end of the spring (or the side which was the end) toward the inside of the hairspring.

Replacing a hairspring and stud is not a difficult job to perform but the finished job often indicates the work of a fine craftsman or that of an

inexperienced apprentice. Simple tools made expressly for a particular job make the performance easier and quicker than the same job done with any tool that happens to be lying on the bench within easy reach. An indifferent attitude toward the work and a mistaken notion that making special tools takes up too much time are detrimental. The making of a studding table or colleting tool will often repay for the time and effort put forth in their manufacture.

FIG. 25.

When the stud has been cleared, the old taper pin should be discarded. It is false economy of time and effort to attempt using the old pin as this will more often than not cause an unsatisfactory job with possible damage to the hairspring.

Replacing the New Stud Pin

A new taper pin about ¼″ long, tapered smooth as illustrated in Figure 26, should be made.

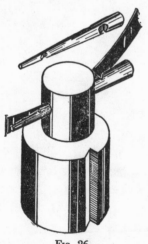

FIG. 26.

. FIG. 27.

To replace the hairspring and to ensure a neat job, place this taper pin into the stud together with a piece of the hairspring to be studded. Push the taper through the hole until the pin and spring become tight but not fast. Mark the points close to the stud on either side of the stud and remove.

When the taper pin is removed, cut two V grooves at the points previously marked as shown in Figure 26. This time, place the hairspring into the stud in the correct direction, making sure of this precaution. Otherwise, the spring may be inserted and fastened in the opposite direction.

After the hairspring is threaded into the stud and resting at the correct vibrating point, the taper pin with its two V grooves is placed into the stud hole so that the pin is outside the coils of the spring and away from it. The pin is inserted so that the thin point faces the end of the hairspring.

The V grooves must face the hairspring as shown in Figure 26.

When this pin is fastened as much as possible by the tweezer by pushing, break off the ends by bending them back as shown in Figure 27. The ends will break off at the grooves previously prepared.

To make the taper pin secure, reverse the process shown in Figure 24.

Taper pins made of hard brass are best suited for this job. Do not try to make the taper pin just the size needed but, rather, make the pin about an inch long, gradually tapered and finished off with a fine file or oilstone slip.

It is then permissible to cut off the last thin $1/4''$. This $1/4''$ length becomes the pin used in the operations just explained. This way it is easier to make the pin and easier to hold. Studding or colleting pins may be bought. These serve many useful purposes.

Summary

1. Studding is that operation in which the hairspring is threaded into the eye of the stud and fastened with a small taper pin.

2. Usually, the stud hole must be cleared of the old hairspring and taper pin.

3. A short, strong tweezers braced against the stud may loosen the old pin.

4. A pinvise may be used to hold the stud during this operation. For specialized work a studding table or vise is used.

5. The use of the balance bridge is not recommended in small watches because it does not permit dexterous use of the fingers.

6. A new pin may be made of brass, tapered to a fine point. The last $1/4''$ is then suitable for the stud.

7. A trial fit of the spring and new stud pin is made.

8. The new stud pin is nicked at the eventual breaking points.

9. The old stud pin should not be used. This is difficult to handle and will not secure the spring firmly.

10. The new pin is pressed into the stud by bracing one point on the thick end of the pin and the other on the opposite side of the stud. Squeezing the stud secures the pin.

QUESTIONS

1. When is the studding operation necessary?
2. Why is it inadvisable to use the balance bridge to hold the stud during the studding operation?
3. What is a studding table?
4. What are two accepted methods of holding the stud for this operation?
5. How long should the new taper pin be made?
6. Why is it inadvisable to use the old taper pin?
7. How is the old pin removed?
8. How is the new pin made?
9. Why is the new pin grooved at either side of the stud during the trial fit?
10. How is the new, shortened pin finally made tight?

XVIII

GENERAL REPAIRS

T he micrometer may be compared to the "C" clamp, Figure 1. If each thread were 1 millimeter apart and the space between the jaws of the clamp could accommodate, say 17 threads, it would be possible to gauge the thickness, in millimeters, of anything that could be held in it.

Shown in Figure 2 is a barrel which gauges 6.00 m/m. We arrive at this figure because we know that 17 threads of the screw will fit between the jaws of the clamp. Since as in Figure 2 there are 11 threads between the barrel and the clamp jaw, the barrel then must occupy the remaining 6 threads of 6 millimeters. However, we must count the threads left in order to arrive at the crude answer. If we could devise some method of marking the threads, we could then read the thickness of the barrel more quickly.

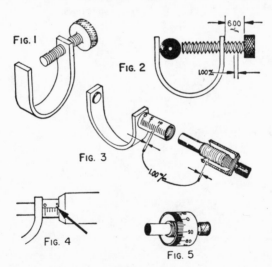

This is accomplished by making the clamp contain a hollow sleeve with inside threads as in Figure 3. The screw now has a thimble over it as shown in the cut-away view. The clamp's sleeve is marked in millimeter divisions so that when the screw is threaded into it, the thimble will cover all but the remaining millimeter markings. The nearest marking to the edge of the thimble is the size of the object in millimeters. Shown in Figure 4, the object gauges 5 millimeters.

In Figure 4, you will notice that the thimble did not stop exactly at the no. 5 but a little bit beyond it. Now, by marking off lines on the thimble, say, dividing the thimble into one hundred markings around it, as shown in Figure 5, the markings in Figure 4 could be told not only in millimeters, but in additional fractions or decimal parts of the millimeter as well.

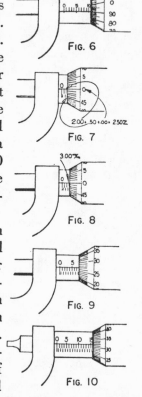

FIG. 6

FIG. 7

FIG. 8

FIG. 9

FIG. 10

In Figure 6, this is shown. In this view, the largest number uncovered by the thimble is read as whole numbers or placed to the left of a decimal. This is shown with the number 10 uncovered. Therefore it should be read as 10. m/m. Notice that there is considerable space between the number 10 and the edge of the thimble. Notice again that the horizontal line on the sleeve coincides with the horizontal markings of 90 on the thimble. Since all markings on the thimble belong to the right of a decimal mark, the thimble marking is read as .90 m/m. If this is added to the 10.00 of the sleeve, we get the total of 10.90 m/m which is the exact reading of this micrometer.

When the thimble is of a large diameter as in Figure 5, 100 decimal markings can be spread around it and yet be read, but when the micrometer is a small one, this would make very difficult reading. Therefore, most bench micrometers which have smaller thimbles have only 50 markings on them as shown in the remaining figures (7-8-9-10). Here the threads on the spindle are only ½ millimeter apart and therefore it requires two turns of the thimble-spindle to move the micrometer one full millimeter. Therefore the sleeve is marked both in full millimeter markings as well as in half markings.

Examine Figure 7. The sleeve is marked off on its horizontal line in full millimeters and the half markings are placed below each and midway between them. The sleeve is marked off in fifty equal divisions. In this figure, the mm thimble has uncovered only the 2 full millimeter markings on the horizontal line but one of the half markings beyond it. Therefore, in addition to the full 2.00 m/m, the .50 of the lower division must be added as well. This brings the interim total of 2.50 m/m. Since the thimble has stopped exactly on the 0 line, nothing is added from the thimble and 2.50 is the final reading. In Figure 8, 3 full markings have been uncovered but nothing of the half mark can be seen. Since the thimble here too has stopped on the zero (0) marking, the exact reading is 3.00 m/m.

In Figure 9, the thimble uncovers 7 full millimeters plus one half additional millimeter. This brings the interim total to 7.50. However since there is still additional space beyond the half marking, we add the thimble marking to the interim total. The reading on the thimble shows .26 m/m. Therefore, this added to the interim total of 7.50 brings the final reading here to 7.76 m/m. See if you can obtain the exact reading of the micrometer in Figure 10. The correct answer will be found at the bottom of this page. In order for you to understand the correct names of the parts of a micrometer, Figure 11 shows a modern micrometer with its principle parts labeled. Practice reading the micrometer and in a short time, it should be very simple to obtain very accurate and quick micrometer readings.

Figure 10 shows a micrometer reading of 16.82 m/m. (16.00 + .50 of the half marking + .32 of the thimble reading)

PART II: HOW TO REMOVE A BROKEN SCREW FROM A BARREL ARBOR

There are many methods of removing a screw from barrel arbors. Before removing the screw or attempting this job, careful consideration should be given to the cause of the screw becoming broken in the arbor. Generally, there are three main causes for screws breaking in arbors. The main cause is attempting to remove a screw with a left thread (one that become loose with clockwise turning) by twisting the screw with the conventional counterclockwise method. This only makes the screw tighter and then the head of the screw snaps off. The second reason is akin to the first, that is, tightening the screw beyond its breaking point. The third, less frequently met, is an inherent weakness in the screw head being too wide and thin with one half of the screw head breaking off while the other half follows after the usual attempt to remove the screw.

After determining the direction of the threads, the barrel arbor is placed into the lathe chuck with the chuck holding the arbor over the shoulder usually occupied by the barrel cover and the square facing outwards. The stump of the screw usually projects slightly from the square of the barrel arbor. Place the T rest close to the tip of the arbor and the screw abut·ment as shown in Figure 7. Place the point of a sharp lathe graver against one of the outer jagged ends of the screw. If the screw is found to loosen by a clockwise motion, place the graver at such a spot on the screw nearer to the front of the lathe and rotate the lathe forward. This will keep the screw stationary while in effect the barrel arbor is twisting itself out of the screw. If the screw is the regular right-threaded type, reverse this process, that is, place the graver at the other end of the screw and reverse the direc-

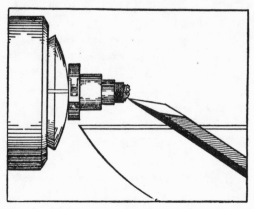

FIG. 7. Removing a broken screw from a barrel arbor.

tion of the lathe. Some watchmakers claim that heating the arbor in oil will facilitate this process but, since the arbor has four bearing surfaces, the temper or polish of these surfaces should not be tampered with.

PART III: DESCRIPTION AND USES OF STAKING TOOL PUNCHES

The usual staking set contains upwards of 80 to 120 punches and an auxiliary of various stumps. To the apprentice as well as to many experienced watchmakers, the use of many of these punches is an endless mystery. Many punches in the staking set were designed for watchmaking

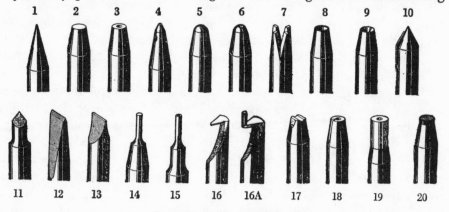

FIG. 8. Staking set punches.

many years ago and, although useless today, are nevertheless still included. On the other hand, many watchmakers are wont to claim that they use only a dozen or two in the entire set. This may be true of their own experiences, but there is a definite use for most of the punches. Other punches used in the staking set are designed to aid the "botcher" in his skulduggery. Many punches that could very well be included in present-day sets to replace outmoded models, have to be purchased as special items.

Since no manual of instructions is included with the sale of these tools, the punches most commonly found in the sets are pictured in Figure 8.

1. Centering Punch

This punch is used to center and lock the staking die to the frame so that the hole selected in the die will be concentric and opposite the punch extended down from the frame arm. This punch is generally very accurately ground and precise, since upon its accuracy rests the exactness of all jobs aligned by it. For this reason it should not be used as a prick punch or for any other operational work.

2. The Flat-Faced Unbored Punch

This is used to replace friction-type jewel bushings or brass bushings in watches or clocks. Occasionally, it may be used to stretch flat steel or brass pieces and to drive out friction pins.

3. Flat-Faced Bored Punch

This punch has many uses and is one of the busiest in the set. It is used chiefly to replace wheels upon pinions, to flush-finish riveting on balance

staffs after they have been riveted to the balance, and in the replacement of friction-type balance staffs.

4. Pivot Punch

This punch might be called by many other names, but its chief use lies in its ability to protect the pivots of arbors from harm while a driving force is exerted upon the adjacent shoulders of the arbors. When a staff

FIG. 9. Cross-hole pivot punch.

FIG. 10. Hole-closing and peening punch.

is desired to be removed without disturbing the pivots, this punch is called into play. It is also used to drive on the friction-type pallet arbors. Some sets come with a selection of these punches and they are also to be had in separate sets. They should be selected for a job with care. The hole in the side is for observation and for removal of a pivot that might become broken and lodged in the hole.

5. Unbored Concave Punch

Used to swedge metal, to close brushing holes, and for repoussé work. Figure 18-10 shows how this punch is made to bear upon a hole (cross-sectioned). It closes by forcing metal down and against the walls of the holes. Its disadvantage is that it leaves the bearing hole walls thin and might prevent their long life.

6. Bored Riveting Punch

It is used to rivet over the countersinks of staffs pinions and arbors. A greater selection of these punches in graduated hole-sizes is to be preferred to many of the outmoded punches in the set.

7. Four-Pronged Riveting Punch

This punch is used to stake wheels onto pinions where great security is desired. Used where plain riveting will not hold, such as on clock pinions.

8. Ring-Holed Punch

This punch, rarely used, is for slight closing of holes in plates when bushings are just a bit too loose.

9. Inverted Cone Punch

To drive in and out concentric pivoted staffs and arbors. It is used to close holes in the pipes of hour hands. (See Figure 11.) This punch may be used also, with extreme caution, in the closing of hairspring collets that have been sprung open. Figure 12 shows a method of removing or driving concentric pivots such as staffs of an alarm clock. It should be noted that the small hollow hole in the punch protects the sharp pivot from damage. These punches should be selected and used with care as they are brittle and may cause damage as well.

10. Pricking or Marking Punch

This punch, with a stronger point than the centering punch, is used to mark centers or to nick. It is shown doing such work in Figure 14.

11. Three-Cornered Punch

This punch has a debatable use. Its use often invites botchwork. An example is shown in Figure 13, where this punch was used to close the hole in the roller. Notice that the metal was nicked and cracked. It furthermore holds only in weak spots.

12. Knife-Edge Punch

Used for parting or cutting thin metal.

13. Peening Punch

Used for stretching metal without cutting or parting the metal in two.

FIG. 11. Mouth tapered punch.   FIG. 12.   FIG. 13. Damaged roller caused by 3 corner punch.   FIG. 14.

14, 15. Special Punches

For removing or knocking out frozen or broken screws and plugs.

16, 16a. Cylinder Punches

For inserting and removing cylinder tampons.

17. Special Punch

For holding cylindrical objects for marking as shown in Figure 14. Also used to loosen frozen tampons from cylinders by placing cylinder across cradle and tapping lightly with flat-faced punch. This tends to widen the diameter of the cylinder, sometimes desired in cylinder escapement work.

18. Roller (Single) Punch

Used to drive single or flat rollers upon staffs.

19, 20. Special Bushing Punches,

Usually with brass nibs to prevent marking the polished bushings.

Stumps

In the modern staking set are many stumps that match the punches in appearance but are designed as dies rather than stakes. However, the staking block frame is designed to permit the use of the punches as stumps. The stumps or inverted punches are very helpful when milled surfaces or recesses are to be worked on.

XIX

TROUBLE SHOOTING

When a watch refuses to go or stops for unknown reasons after repair and overhauling, the source of trouble must be tracked down and eliminated.

Detecting trouble in the watch is often a matter of observing symptoms and by a process of elimination arriving at the cause of the trouble. Some irregularities will not stop the watch but cause other inconveniences such as interference with the hand-setting mechanism or the winding parts. Therefore, the search for causes of malfunctions must start at the point which it affects. Should the watch refuse to go, the balance is inspected first. Below is a trouble-shooting chart, listed according to divisions, and in a logical searching sequence.

If the watch should stop, first examine the dial for the following:

Causes	*Corrections*
Hands catch	Provide proper clearance
Second hand binds on dial . . .	Provide for clearance
Minute hand rubs crystal	Lower minute hand
Dial loose, rubs hour wheel post .	Center dial and tighten dial screws
Second hand (fourth wheel) pivot bent	Straighten and burnish pivot

If the above symptoms are eliminated as being the cause of stoppage, open the watch case, remove the movement, and examine the balance. With a fine brush, push the balance so that it turns. If—

268

The balance does not turn

Causes	Corrections
Broken balance staff	Replace the staff
Bent balance staff pivot	Straighten and burnish
Balance touches the center wheel .	True balance or change staff
Balance lies on pallet bridge . .	Examine lower pivot, cap and hole jewels
Balance overbanked	Examine causes for overbanking
Loosened timing or balance screw .	Replace and make secure
Hairspring rubbing on balance arms	Level and center hairspring
Roller jewel binding in fork slot .	Replace with proper fitting jewel pin
Roller table rubbing on pallet fork	Examine lower pivot and jewels
End of hairspring sticking into balance	Straighten hairspring
Loose balance stud	Raise the stud and secure
Regulator touches balance . . .	Examine regulator if bent; true balance
Tangled hairspring; badly magnetized spring	Untangle spring; demagnetize

Balance turns but stops when watch is turned dial-side up

Causes	Corrections
Upper cap jewel loose or broken .	Fasten jewel or replace
Bent or broken pivot	Straighten and replace
Lower pivot end mushroomed . .	Burnish pivot
Balance touches regulator . . .	Examine upper pivots and jewels
Balance touches center wheel . .	Examine upper pivot and jewels; true balance
Balance rests on hairspring . . .	Examine endshake, upper pivot and jewels, level hairspring

The balance turns but soon stops

Causes	Corrections
Dirty pivots and jewels	Overhaul and clean
Bent balance	True balance in calipers
Magnetized hairspring	Demagnetize
Bent hairspring	True hairspring

Balance screw washer rubs pallet
   bridge or center wheel . . . .   Replace with correct sized washer,
                               poise and regulate
Bent pivots . . . .   Straighten and burnish
Broken jewels . . . .   Replace jewels
Balance scraping pallet bridge . .   Examine lower pivots and jewels

The balance has a good motion but the pallet is motionless

Causes   *Corrections*

Broken or loose roller jewel . . .   Replace roller jewel

Balance and pallet move but both soon stop

Causes   *Corrections*

Chipped pallet jewels . . . . .   Replace pallet jewels
Roller jewel binding in pallet . .   Replace with thinner jewel
No endshake in pallet . . . . .   Adjust endshake
Gummed pallet arbor pivots . . .   Peg out pallet hole jewels
Loose roller jewel . . . . . .   Shellac roller jewel
Roller jewel out of upright . . .   Straighten roller jewel
Gummed oil in all jewels . . . .   Overhaul movement
Balance pivots bent . . . . .   Straighten pivots and burnish
Cracked hole jewels . . . . .   Replace jewels

If the balance and the pallet move but soon stop because the escape wheel is motionless, the trouble is back farther in the movement. There may be some obstruction in the train. One method to discover the trouble spot in the train is to grasp the center wheel by its spokes while the pallet and bridge are still in the watch and move the center wheel back and forth. If—

The center wheel cannot be moved at all

Causes   *Corrections*

Loose screw jammed between pinion
   leaves and barrel . . . . .   Release power and remove obstruction

Stem pilot pin too long, sticking into
   center pinion . . . .   Shorten pilot pin
Screw jammed between center wheel
   and third wheel pinion . . . .   Release power and remove obstruction

Crown wheel screw too long and
   sticking into center wheel teeth .   Change or shorten screw
Hands caught or dial train jammed .   Release hands or repair dial train

The center wheel moves the third wheel but the fourth wheel remains motionless

Causes	*Corrections*
Second hand touches dial	Provide clearance for hand
Fourth wheel pivots bent	Straighten pivots and burnish
Obstruction between fourth wheel teeth and escape wheel pinion .	Remove obstruction
No endshake in the fourth wheel .	Provide for endshake
Cracked or broken fourth jewels .	Replace jewels

The center, third and fourth wheels move but the escape wheel will not budge

Causes	*Corrections*
Bent escape pinion pivots	Straighten pivots and burnish
Loose or broken escape pivot cap jewels	Fasten or replace cap jewels
Loose pallet jewel jammed under escape wheel	Refit pallet jewel
Lower escape cap jewel screw too long, and butts against the escape wheel	Exchange for shorter jewel screw, probably placed elsewhere in watch
Loose escape wheel bridge . . .	Fasten bridge screws, test endshake

If the balance, escapement and train move but stop after a short time, wind the watch fully. If—

The watch picks up motion

Causes	*Corrections*
Mainspring is "set"	Replace mainspring
Watch needs overhauling	Clean and oil movement
Barrel cover may be loose . . .	Replace and secure barrel cover

The watch cannot be fully wound

Causes	Corrections
Broken mainspring	Replace mainspring
Broken barrel hook	Repair hook
Mainspring endpiece not sharp enough to catch on barrel hook .	File under part of tongue of endpiece
Barrel arbor disconnected with inner coil of mainspring	Curve inner coil to grasp barrel arbor hook
Barrel cover disconnected from barrel	Replace and secure

The ratchet wheel unwinds

Causes	Corrections
Broken clickspring	Replace clickspring
Broken click	Replace or repair click
Click too tight on boss	Adjust or repair click or boss
Improper clickscrew	Replace with correct screw
Loose clickscrew	Fasten clickscrew
Loose ratchet screw	Secure ratchet screw
Broken crown wheel or ratchet teeth	Replace wheel

Winding skips or scratches

Causes	Corrections
Worn winding pinion or winding wheels	Replace with new wheels
Loose crown wheel washer . . .	Replace with correct fitting washer
Broken ratchet, crown and winding pinion teeth	Replace and search for missing teeth to prevent their clogging other parts
Stem shoulder too long, preventing clutch wheel from fully engaging winding pinion	Place stem on lathe; turn back winding pinion shoulder
Crown wheel screw too long, possibly interchanged with ratchet screw	Change for shorter screw

Stem too loose in winding pinion . Change stem or winding pinion
Stem bearing worn . . . . . Re-bush or fit with oversize stem

Watch is hard to wind

Causes	Corrections
No oil in crown wheel washer . .	Lubricate washer
Rusted winding wheels	Clean and lubricate
Crown is worn or too small . . .	Replace with correct-size crown
Stem and crown binding in case . .	File out case hole
Stem setting lever slot too narrow .	Adjust stem slot
Barrel cover loose	Replace and secure barrel cover
Ratchet wheel rubs on barrel bridge	Tighten bridge screws or check barrel for upright

The crown unscrews

Causes	Corrections
Threads stripped	Replace crown or stem
Stem rusted	Remove and clean
Stem binds in case	Provide for clearance
Crown not fastened properly . . .	Tighten crown
Dry winding pinion, clutch wheel or stem square	Lubricate winding parts

Stem pulls out of movement

Causes	Corrections
Setting lever screw loose	Tighten setting lever screw
Setting lever pin worn or too short .	Replace setting lever
Stem (plate and bridge) bushing worn	Rebush or replace with oversize stem
Stem hubs too thin	Replace with stem with thicker hubs

Watch does not wind or set

Causes	Corrections
Crown threads stripped	Replace crown
Broken clutch lever spring . . .	Replace spring
Clutch lever out of clutch wheel .	Insert clutch lever into clutch wheel, secure setting bridge screws, examine pilot pin of stem

Loose barrel bridge . . . . .   Secure bridge screws

Stem broken off at square . . .   Remove broken parts and replace

Watch is hard to set

Causes	*Corrections*
Crown wheel screw too long . . .	Check for interchanged ratchet or crown screws, replace with shorter screw
Set parts and winding wheels rusted	Clean and lubricate
Cannon pinion too tight	Loosen cannon pinion with smooth taper pin
Burred dial train wheel teeth . .	Repair teeth of these wheels or replace
Hour wheel jammed between dial and minute wheel pinion . . .	Replace hour wheel or provide dial washer if necessary

Watch will wind but not set

Causes	*Corrections*
Setting lever loose or bent . . .	Secure setting lever screw
Pilot pin of stem broken	Remove broken part and replace with new stem
Stem square too short	Replace with proper fitting stem
Lower teeth on clutch wheel, or intermediate wheel broken . . .	Replace clutch wheel or intermediate wheel, overhaul watch to remove broken teeth
Minute wheel teeth broken . . .	Replace as above
Worn lower pilot pin bushing . .	Rebush pilot pin bearing
Loose setting lever cap or bridge .	Secure setting lever bridge screws
Center post broken	Replace center pinion or new wheel complete

Minute hand can be set but hour hand does not move

Causes	*Corrections*
Broken minute wheel pinion, broken hour wheel teeth	Replace these parts
Hour wheel not engaged with minute wheel pinion	Replace and provide dial washer if necessary

Hour hand moves but minute hand does not

Causes	*Corrections*
Broken cannon pinion leaves . . .	Replace canon pinion
Cannon pinion rides up above minute wheel	Tighten cannon pinion accurately

Watch will set but not wind

Causes	*Corrections*
Broken clutch lever spring . . .	Replace spring
Broken mainspring	Replace mainspring
Loose crown wheel	Tighten crown wheel screw
Broken winding pinion	Replace winding pinion
Clutch wheel upper teeth broken or worn	Replace clutch wheel or winding pinion

Watch runs, winding and setting satisfactory but hands do not move

Causes	*Corrections*
Loose cannon pinion	Tighten cannon pinion
Broken center post	Replace center pinion or center wheel completely

Watch stops every six seconds

Causes	*Corrections*
Broken or bent teeth or escape wheel out of round	Repair or replace escape wheel
Broken or bent escape pinion leaf or sand particle in one of pinion leaves	Repair or replace wheel and pinion. Remove obstruction

Watch stops once a minute

Causes	*Corrections*
Second hand touches the dial . .	Straighten hand
Fourth wheel pivot bent	Straighten pivot
Bent or broken fourth wheel tooth .	Repair or replace wheel
Loose dial	Secure dial snugly to plate and fasten

Second hand socket rubs dial hole .   Center dial hole and secure dial (Examine for bent fourth wheel pivot)

Second hand socket rubs hole jewel .   Lift up second hand or shorten socket

Watch stops once in five to eight minutes

Causes	*Corrections*
Bent or broken third wheel teeth	Repair or replace third wheel
Broken or rusted third pinion leaves	Clean out rust or replace pinion

Watch stops once an hour

Causes	*Corrections*
Broken or bent center wheel teeth .	Replace center wheel
Broken or clogged center pinion .	Remove obstruction or replace broken pinion
Clogged or broken cannon pinion leaves	Replace or clean cannon pinion
Hands catch on dial, glass or scrape each other	Provide proper clearance between bands

Watch stops once every few hours

Causes	*Corrections*
Bent, broken or clogged, barrel teeth	Repair or replace barrel
Broken mainspring endpiece or mainspring broken near endpiece	Repair or replace mainspring
Mainspring endpiece slips off hook .	Repair endpiece
Barrel hook broken	Repair barrel hook
Loose barrel cover	Replace cover and secure
Broken or squashed minute wheel or dial train teeth	Repair or replace dial train wheels

Watch stops when in case but starts when taken out of the case

Causes	*Corrections*
Balance has insufficient endshake .	Test endshake and provide for it
Case too tight	Test case and ream it if necessary
Case back too thin	Change case
Timing screws out too far . . .	Turn in these screws, poise and re-regulate

Watch only goes when in case (Negative Setting)

Such a watch setting mechanism jumps into the setting position when out of the case. This causes the dial train to carry the extra load of turning the winding arbor plus the setting wheel. This undue strain may cause the watch to stop. If the watch movement must be timed out of the case, the winding arbor will have to remain in the winding position. Most watches with such windings have a set screw or special lever which is adjusted to keep them in the winding position.

Self-Winding Watches

Watch does not store sufficient power

Causes	Corrections
Mainspring slips prematurely . .	Straighten brake spring
Oscillating weight scrapes case back	Replace axle bushing or jewel. Tighten movement in case, test thickness of gasket
Retaining click does not grip . .	Stiffen clickspring, examine winding wheel teeth
Breguet pinion teeth worn . . .	Replace pinion
Reverser arm sluggish	Clean reverser and bearing holes, oil pivots
Pawl bearing wheel teeth (ratchet shaped) worn	Replace wheel

Watch rebound due to excessive winding

Mainspring barrel dry	Lubricate mainspring barrel
Brake spring grips too tightly . .	Lubricate brake spring with graphite or molybdenum disulphide
Brake springs grips tightly due to stiffness of spring	Provide more curvature to brake spring

Watch stops after assembling shockproofing device

Upper and lower cap jewels interchanged	Place thinner cap jewel on dial side
Safety roller counterbore squashed in replacement	Replace with new, genuine roller
Burrs on shockproof bushing tube (for pivot)	Replace bushing or remove burrs
Retaining spring not properly locked in place	Provide proper locking for spring
Bushing loose in plate or bridge .	Push in locking bolt securely
Retaining spring bent or cracked .	Replace with genuine part

XX

SERVICING WATCH CRYSTALS

With the modern need for water-resistant watches, the types and fitting of watch crystals have achieved great importance for the watchmaker. The crowns and cases for such watches have been reinforced. In many instances, the case back has been incorporated with the frame to become a one-piece case with the movement removed from the crystal side. This means that the crystal must be designed so that it can be removed and inserted as often as the movement needs to be serviced, without the crystal becoming marred, losing its resiliency or water-resistant qualities. In such watches, the crown and stem snap on and off.

As a result, crystals are made not only thicker, but also with considerable resiliency. In this way, they can be temporarily constricted and then relaxed in their bezels again and again, and still exert a water-sealing pressure inside the bezel groove.

Another type of crystal now becoming popular for water-resistant watches is the armored crystal. This type has a supporting metal ring on its inside rim. The purpose of this reinforcement is to strengthen resistance to moisture entry as well as to shrinkage due to the aging process in almost all types of plastic crystals. This reinforcement excludes the usual mechanical constriction to seat the crystal into its bezel; thus, special tools are required.

The new types of watch cases and crystals could not be marketed until specialized tools to service them were designed and made available to the watchmaker. While some of these tools are currently on the market, often, they are used without full knowledge of their application.

Lifting Tool

The most common type of crystal tool is the crystal constrictor or lifter shown in Figure 1. In essence, it is a chuck with about 15 prongs which grip the periphery of the crystal. Tightening the chuck constricts the crystal, doming it somewhat at its center. This reduction in diameter allows it to loosen in its bezel seating, and, thus, it is easily removed.

Such a tool allows easy adjustments to the hands and dial, instead of using the older method of removing the case back, case screws, stem and crown in order to remove the movement, merely to get to the dial and hands.

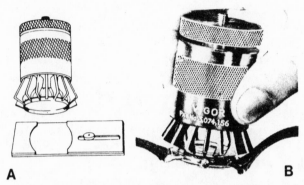

A   **B**

Fig. 1. Vigor tool has 15 prongs which constrict and lift a watch crystal from its bezel.

With the constrictor, the crystal is grasped by the prongs and removed. It may remain in the jaws of the constrictor for a short while. When it has to be replaced, the crystal is carefully seated in its bezel, twisted slightly to make certain that all of the crystal is within the bezel groove, and the upper chuck knob is loosened, allowing the crystal to spread out and resume its tight place in the bezel. The crystal should not remain in the tool too long as this might cause "crazing."

When a new crystal is fitted, a little more care must be taken. If a crystal too small in diameter or one just the same size as the bezel is chosen, it will actually become loose in time. In the meantime, it will not have sufficient outward pressure against the bezel groove to stop the entry of moisture or water. Therefore, the watchmaker should use a gauge to measure the inside diameter of the bezel. A crystal with an outer diameter 0.50mm larger than the bezel's inner diameter should be chosen.

The larger diameter allows the crystal to be constricted and fitted to the bezel and then relaxed so that its outward force forms an impervious seal. On the other hand, a crystal with a diameter more than 0.50mm larger than the bezel's can still be constricted with these tools with a bit more force of the tool chuck. However, when such a crystal is fitted, its internal molecular structure is always under severe stress. This causes the molecular structure to break down, and the crystal ages prematurely. This is shown by "crazing," the appearance of numerous fissures and cracks accompanied by a discoloration of the plastic. Such a crystal is not water- or moisture-resistant. For the same reason, a crystal should not remain in the tool overnight.

Crystal Edges

When a new crystal of the proper diameter is chosen, the jaws of the crystal lifter must grasp the periphery a distance up from the bottom so that some of the edge of the crystal projects out of the chuck to provide a lead-fit into the bezel. Extreme care must be taken not to have the prongs grasp the crystal on any portion of its edge that will be in the bezel. If the edge is grasped, the sharp prongs will make tiny dents which allow moisture and water to enter the watch. Any crystal with scratches or dents on this section is not water-resistant.

To provide a means by which the constrictor-lifter prongs grasp the crystal at a safe height, the tool shown in Figure 1 has a platform on which to place the crystal. The sliding plate then moves snugly against the crystal and is secured by the knurled nut. Then the lifter is placed over the crystal, with its jaws resting on the platform. Since the edge of

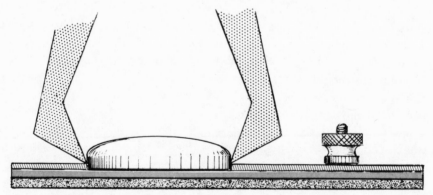

FIG. 2. To prevent "crazing" the tool grasps the crystal a distance up from its edge equal to the thickness of the sliding plate into which it is fitted.

the crystal is below the sliding plate, the tool's jaws grasp the crystal a distance up from the edge equal to the thickness of the plate (Figure 2).

The tool is then tightened, constricting the crystal just enough to enter the bezel. The top knob is loosened gradually, making certain that the crystal remains firmly seated in the bezel. A slight twist of the tool itself indicates whether the crystal is still seated in its bezel. Then the tool is loosened and removed. Most watchmakers press on the crystal with the watch between thumb and forefinger as an added precaution that the crystal is firmly seated.

Another crystal lifter and inserter with some unusual features is shown in Figure 3. One of its features is that it can easily be adjusted to grasp the crystal at five different heights from the bottom edge of the crystal: 0.01 in., 0.25mm; 0.02 in., 0.50mm; 0.03 in., 0.75mm; 0.04 in., 1.00mm; 0.05 in., 1.25mm. This, it is claimed, is an important feature because the depth of the bezel's crystal groove varies from one watch case to another.

The tool has 15 prongs and operates like most other crystal lifters. When the crystal must be removed, the tool is placed over it and the prongs tightened until the suspended watch drops a short distance to the bench top. If the crystal is removed for a short-term access to the movement, it need not be removed from the tool. Replacing the crystal is done in the manner previously described.

Choice of Height

If a new crystal is needed, its diameter should be 0.50mm larger than the bezel. Next, the watchmaker must decide at which height the prongs are to grasp the crystal edge. As mentioned earlier, five heights are possible. Notice in Figures 3 and 4 that in the fluted edges of this tool are the numbers 10, 20, 30, 40, 50, representing thousandths of an inch. Notice, also, that below these numbers on the bottom edge of the tool are a series of four studs, one higher than the next, the highest under the number 50. Under number 10 the surface is flush. There are three sets of such studs.

The three-pillared base of the tool is made to fit into the flutes around the tool edge. The flute in which the pillars are fitted determines the height at which the crystal is grasped. In Figure 4, the height is 50 thousandths of an inch or 1.25mm.

The crystal is placed so that the three stepped pillars fit into the numbered flute designating the required height at which the prongs are to grasp the crystal's edge.

Pressing the knurled button on top of the tool, positions the coned-cup on the crystal (Figure 5). Here the stepped pillar rests in the flute marked 40. The stud on the bottom of the tool rests on the step of the pillar. This raises the tool's prongs off the base a distance equal to the height of the studs. In this case, it is 40 thousandths of an inch, 1mm.

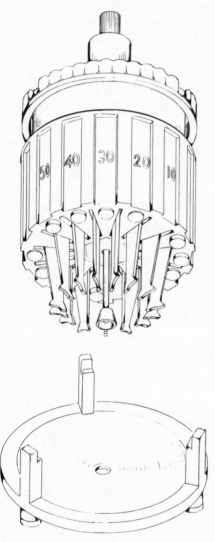

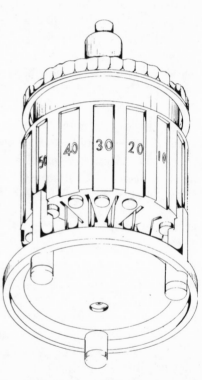

Fig. 4. Numbers on the fluted columns of the tool tell the height at which the crystal can be grasped. Here, the height selected is 50 thousandths of an inch.

Fig. 3. Standard Unbreakable Crystals makes a lifter which adjusts to grasp the crystal at five different heights.

Thus, the prongs are shown grasping the crystal just one millimeter up from the bottom of the crystal. If the studs numbered 50 rested on the pillar-step, it would raise the prongs 50 thousandths of an inch from the crystal's bottom, as shown in Figures 4 and 6.

In Figure 7 the tool is shown resting on the pillars without any stud. This provides the minimum clearance between tool prongs and crystal bottom, 10 thousandths of an inch or .25mm. The extra studs and prongs in this illustration have been omitted for clarity.

Before fitting any crystal, it is a good idea to use a pointed pegwood stick to clean out dirt in the bezel groove. This tool can also be used for this purpose by fitting the outside portion of the prongs to the desired size and twisting in the bezel until the dirt is removed.

Certain crystals are now being made with profiles resembling Figure 8. These are stepped so that any pronged crystal lifter will automatically grasp the crystal at the correct height for insertion.

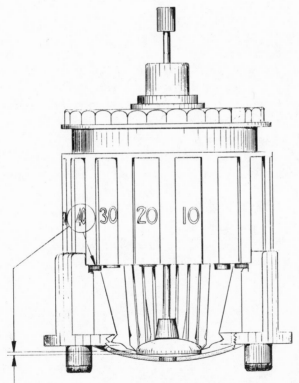

FIG. 5. The knurled button on top of the tool positions the coned cup on the crystal. The studs are 40 thousandths of an inch from the crystal edge.

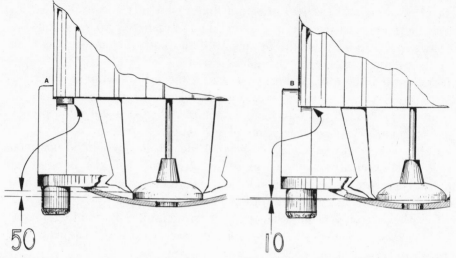

Fig. 6 (left) and Fig. 7 (right). The height of the studs is shown at 50 and at 10 thousandths of an inch.

Replacing Armored Crystals

With demands on watches to pass severe tests of water resistance, many new watches are being housed in watch cases whose crystals are reinforced with a metal ring on the inside base. These are called "armored crystals," "tension-ring crystals" or simply "reinforced crystals." This ring not only strengthens the crystal, but also aids in exerting an outward force against the bezel wall, creating a more effective barrier to moisture, especially in watches subjected to the great pressures of the deep.

An armored crystal also lengthens the operative life of the crystal. Practically all organic crystals can be expected to shrink with age. Thus, they become smaller in diameter and lose their water-tightness. With an inner armored ring, shrinking is inhibited somewhat.

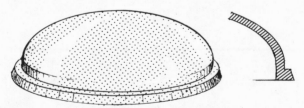

Fig. 8. A stepped crystal enables any lifter to grasp it at the correct height.

Most armored crystals come with straight walls and do not have a flared out edge. These crystals cannot be wedged into an undercut recess but must remain secured, flush against the bezel wall, by virtue of being slightly larger than the bezel. They must be pressed down into a recess which is usually 0.10mm to 0.16mm smaller than the crystal. Some makers of these crystals warn against using a crystal whose diameter is more than 0.16mm larger than the bezel. Too large a crystal may crack or craze.

In most crystals, the tension ring is split to allow contraction without crimping the ring. Such a crystal and its split tension ring are shown in Figure 9. Many newer types are being introduced with solid, unsplit rings as an integral part of the crystal. These crystals must be chosen with even greater care for size; if one too large for its bezel is fitted, the tension ring will buckle, destroying its purpose and effectiveness and possibly cracking the crystal.

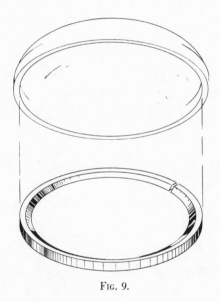

FIG. 9.

As shown in Figure 9, tension rings are highly polished and beveled on their inner surfaces presenting a pleasing appearance and acting as a frame reflector.

Crystal Presses

To push these crystals into place, some force is required. To do it precisely and without damage, the watchmaker should use a special

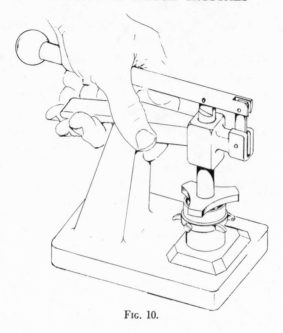

FIG. 10.

press with "chucks" or "fixtures." A number of crystal presses are available, each claiming certain merits. These hand presses hold cylindrical crystal fixtures of varying diameters which can be interchanged with pushers, rests and stumps.

Figure 10 shows a typical armored crystal inserter and will be used to illustrate some of the recommended methods of inserting an armored crystal. Most crystal presses come with two sets of pushers or resting fixtures. Figures 11, 12 and 13 show how they are used and will be explained in detail later.

Each fixture has a top and bottom with the same characteristics. For example, the fixtures with beveled insides, Figures 11B and 13E, are designed to accommodate the rounded edge of the crystal. The smallest fixture in one of the sets accommodates crystals from 16.0mm to 16.9mm on one side and from 17.0mm to 17.9mm on its opposite side. The fixtures are graduated in one millimeter increases so that a set of 10 beveled fixtures encompasses a range of 20 millimeters, up to and including 35.9mm.

When inserting an armored crystal into the case bezel, the jeweler should choose a crystal whose outer diameter does not exceed the manufacturer's specifications. The next consideration is whether the watch case is a one-piece or a two-piece type. In a two-piece case the back can be removed by unscrewing it or snapping it off. If the movement is in the case and one does not wish to remove the movement or unscrew the back, then the unit, regardless of the type, should be treated as a one-piece case.

One-Piece Case

If the crystal is to be inserted while the movement is encased and the case back secured, the watchmaker must choose the case rest or bottom platform with care. If a solid, flat base is used with the case back resting on it, any pressure on the case back may cause the back to buckle, warp or bend inward, breaking the watch parts inside.

To insert an armored crystal while the movement is encased, first select a base fixture with a counterbore to accommodate the back rim of the case so that the fixture not only supports the strongest part of the case

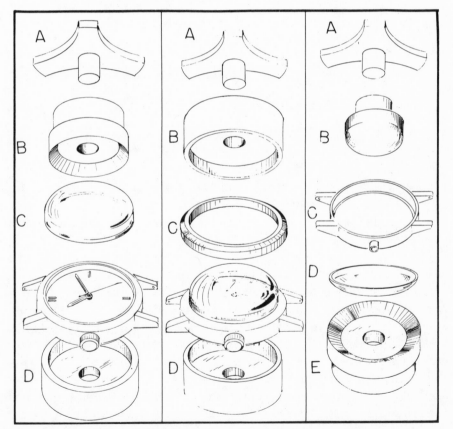

Fig. 11 (left), Fig. 12 (center) and Fig. 13 (right) show three sets of crystal pushers and resting fixtures for different uses of the crystal press.

back, but also fits close enough to serve as a centering chuck as well. Fit this fixture over the base's center post. Next, select a beveled fixture whose slanted sides just touch the curved edge of the crystal, as in Figure 14.

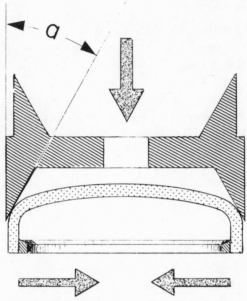

Fig. 14 shows how the angle of its beveled fixture will contract the crystal's diameter for easy insertion.

When the set comes with the fixtures marked in millimeters or sizes, they should be used according to the diameter of the crystal chosen. The beveled fixture is then attached to the pusher post. The watch and fixture should then appear as in Figures 10 and 15.

If a squeezing pressure is applied to the crystal, as in Figure 10, better control is attained than by using one's own weight. All presses have a spiral return spring to lift the handle when hand pressure is relaxed. (For clarity the spring has been omitted in the drawings.) The purpose of the three-winged upper plate (Figures 10, 11A) is greater stability for the fixtures. It also allows more dexterous finger manipulation when fitting the fixtures.

To insert an armored crystal in a case where the movement and case back have been removed, rest the case on a fixture that fits inside the case. For the insertion, a bottom fixture of the type shown in Figure 11D may also be used. Here the straight inside fixture supports the case, providing that it fits closely for proper alignment with the upper fixture.

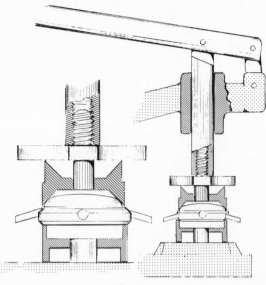

FIG. 15.

Uses of the Crystal Press

Figures 11, 12 and 13 show three different uses of the crystal press. Figure 11A shows the three-winged top alignment plate; B is the beveled fixture with slanted sides to hold the rounded edge of the crystal C to be pushed down inside the straight walls of the watch case's bezel. D is the straight counterbore fixture that holds and supports the outer rim of the case back.

In Figure 12, the crystal has no tension rings. Instead, the crystal fits over a rim which is part of the watch case. A special outer rim or frame C must be pressed on to the outside of the crystal to tighten it and seal the water-resistant watch. The diameter of the upper fixture B must allow a slightly loose fit over the crystal but must make solid contact with the bezel frame C.

Figure 13 shows how the watchmaker can use the same tool to insert ordinary plastic crystals. Here, the rubber press B, of the proper diameter, is fitted to the upper fixture. The crystal rests on a beveled base whose diameter is slightly smaller than the crystal's diameter. The plunger goes through the case bezel. As it contacts the crystal, it presses slightly, causing the crystal to bell downwards, reducing the diameter until the crystal can be eased into the bezel. When the pressure is relaxed, the crystal expands outward to its normal shape and size, seated firmly in the bezel.

APPENDICES

I: A Dictionary of Trade Terms

While this dictionary of trade terms includes the basic terms and definitions used in this trade, many more terms are defined in the text. The comprehensive index at the rear of the book lists these numerous other terms, phrases and tools for which adequate definitions can be found in the text.

Abrasive: A substance used for grinding and polishing. In watchmaking these are generally oilstone, emery, carborundum, lavigated aluminum oxide.

Accumulator: Part of an electric watch which takes the place of a battery and can be recharged to accumulate energy.

Acetone: A chemical liquid used to dissolve celluloids; used in crystal cement, and as a dehydrator after cleaning watch parts although not well recommended as a rinse because of its violent inflammability.

Alum: A whitish mineral salt used by watchmakers to dissolve steel screws that have broken off in plates of *other* metals such as brass or copper.

Amplitude: The amount of arc or swing of the balance. This can also be measured by certain types of oscilloscopes.

Analogue: Terms used to denote any timepiece with dial and hands as opposed to "Digital display."

Ancre: The Swiss term for the pallets.

Annealing: The act of heating and slowly cooling a metal or substance to render it softer or to relieve internal stresses.

Arabic figures: Figures on a dial, such as 1, 2, 3, as opposd to *Roman numerals*, such as I, II, V, IX.

Arbor: The axle of a wheel or a shaft that turns in a bearing; commonly referred to as the barrel arbor, pallet arbor, winding arbor.

Arkansas stone: A white marble-like stone used in various shapes and sizes as a grinding stone to sharpen gravers and tools. Used in powdered form as an abrasive with grinding laps.

Automatic: A self-winding watch using a rotor or oscillating weight.

A.W.I.: The American Watchmakers Institute, national watchmakers organization.

Balloon chuck: A lathe chuck with a hollow round box end. A tiny hole is drilled in the front center of the box. The balance wheel pivot is made to emerge from this hole while a pump center punch keeps the balance in place. The protruding pivot may then be ground or polished to size.

Banking pin: Two pins that limit the motion of the pallet fork. They also control the amount of slide of the pallet jewels.

Barrel arbor: The axle of the barrel around which the mainspring is coiled.

Barrel hook: A hook or slot in the inside of the barrel wall upon which the last coil of mainspring is attached.

Beat: The tick of the watch or one swing of the balance wheel. *In Beat:* when one beat or *vibration* of the balance receives its impulse at the same distance from the line of centers as the other.

Beeswax: A tough, yellow-brown wax used as a temporary adhesive by watchmakers.

Bell metal: An alloy of 4 parts copper and 1 part tin, sometimes also containing small quantities of zinc, lead, iron, silver, bismuth, and antimony. Used for metal laps and slips to obtain a high polish on steel.

Benzene: A coal tar product used as a cleaner and rinse in watch and clock cleaning. Highly volatile and inflammable. (Benzol) (C_6H_6). More expensive than *benzine*.

Benzine: A colorless inflammable, volatile liquid obtained from petroleum by fractional distillation, used as a rinse in watch and clock cleaning. Not as volatile as *benzene*.

Beryllium: A metal used in minute quantities with nickel and steel to produce nonmagnetic, noncorrosive balances and springs capable of good temperature adjustments.

Bezel: The top part of a watch case that contains the crystal.

Bimetallic: Made up of two metals. In watchmaking, this refers to the split balance having a brass rim and steel frame.

Bluing: To change the color of polished steel by heating it to approximately 540° F.

Bob: The ball or weight at the lower end of a pendulum.

Boiling-out pan: A copper pan used to boil out broken steel screws (see alum) or to dissolve cement; to boil parts in a weak acid solution; to temper steel in boiling oil.

Boxwood: A hard, close-grained wood used for pegwood or polished laps.

Brace: The hook or connection attached to the outer end of the mainspring.

Brake spring: The extension to the regular mainspring in an automatic watch which supplies the slip-clutch effect to prevent tight winding.

Breguet: Abraham Louis, 1747-1823, horological genius and inventor. The name applied to the type of hairspring which has its last outer coil raised above the body of the spring and curved inwards.

Bridge: The upper plates in a watch movement that contain the bearings for the wheel pivots and have pillars at both ends, such as the train or barrel bridges.

Bridle: Another term for the brake spring. Also called *slip-clutch, safety for mainspring, safety-spring, slip-spring, slipping attachment.*

Broach: A tapered steel tool, with flat cutting edges used to enlarge holes already drilled.

Buffstick: A stick with a strip of chamois glued to it to remove fingermarks from polished steel during final inspection.

Burnisher: A hard polished piece of steel used to polish softer metals by rubbing it along the surface to be finished.

Butting: The action of two wheels or a wheel and pinion improperly matched or distorted wherein their teeth butt each other instead of enmeshing perfectly.

Calibre: The size or factory number of a watch movement.

Caliper: An instrument with two adjustable curved legs, used to measure diameters.

Caliper (balance): A tool shaped like a caliper but with a hole in the top of each leg to hold the balance so that it may be observed for true running in the flat and round.

Cam: A small flat piece used to transfer circular motion into back-and-forth motion to a lever or other contacted piece.

Cannon pinion: A thin, steel tube with pinion leaves at its lower end and usually carrying the minute hand at its upper end.

Cap jewel: The flat solid jewel upon which rests the pivot end. Also called the *endstone.*

Carbide: The type of cemented carbide tools used as gravers to cut very hard metals. Excellent in cutting out balance hubs.

Carbon Tetrachloride (CCL₄): A non-inflammable solvent of good cleaning qualities but highly toxic.

Carborundum: A carbon silicon substance used as an abrasive.

Castle wheel: The clutch wheel.

Cavitation: The process during ultrasonic cleaning in which minute bubbles are created and collapsed to dislodge dirt from parts to be cleaned.

Cement (jewelers'): Stick shellac or sealing wax.

Chain (fusee): A miniature "bicycle" chain connecting the barrel and fusee of English watches and chronometers.

Charged: To have imbedded in a lap of metal or wood, particles of abrasive or polishing powders.

Chronograph: A watch with hour and minute hand and a center sweep-second hand which can be controlled by a special button. A watch where the second hand may be started, stopped and made to return to zero.

Chronometer: A watch or clock adjusted to keep exact time. Usually a large, mounted watch with the detached spring detent escapement.

Circular escapement: The lever escapement wherein the center of the pallet's lifting surface is planted on a circle whose center is the pallet arbor. Found mostly in American watches.

Cleaning solution: A dirt, tarnish- and grease-dissolving liquid composed mostly of ammonia, oleic acid and water.

Click: The pawl used to prevent the ratchet wheel from turning back after the mainspring has been wound.

Clockwise: The direction of circular motion going in the same direction as the hands of a clock; circling from horizontal left, upward around and down toward lower right.

Club tooth: The boot-like shape of modern lever escape wheel teeth.

Clutch wheel: The cylindrical winding wheel with ratchet teeth on its upper end, a slot-neck in the middle, and curved teeth on its lower end. Also called the *castle wheel.*

Cock: An overhanging support for a bearing such as the balance bridge. A bridge having a support at one end only.

Compensating balance: A bimetallic balance of brass and steel split near the arms and constructed so that its effective diameter will contract or expand in temperature changes to compensate for these changes to itself and to its hairspring.

Concave: A surface curved inward as the inside of a bowl or the oil cup of a jewel.

Conical pivot: A pivot which curves back into the main body of its arbor, such as those used with cap jewels. (Balance staff pivots.)

Convex: A domed surface as the top of a curved watch glass.

Corner safety: The distance between the fork slot corner and the roller jewel when they are opposite each other.

Counterpoise: To bring into balance by placing opposite weights to balance or poise. A weight that balances another weight.

Countershaft: An adjunct to the lathe that receives the power from the motor and by belting transmits this to the lathe spindle.

Countersink: A chamfered or concave cut.

Crescent: The notch in the safety roller cut to permit the passage of the pallet fork guard finger.

Crocus: A red-brown powder used for quick polishing, slightly coarser than jewelers' rouge.

Crown: The winding button on top of the stem.

Curb pins: The two regulator pins almost pinching the hairspring.

Curve test: A test similar to the corner safety test but conducted along the curves of the fork horns.

Cyanide (potassium): A poisonous, white crystalline substance dissolved in water to brighten tarnished metals.

Dead beat: An escapement in which the escape wheel remains motionless after it becomes locked upon the pallets as opposed to the recoil escapement. Generally, as referred to the Graham escapement used in precision clocks.

Dehydrate: To deprive of water. In watchmaking to dry or rinse off the cleaning solution and dry the watch parts.

Demagnetizer: An electric coil connected to an A.C. power supply through which a magnetized movement or part is passed to diminish its magnetized condition. Newer models of this device use capacitors, electronically for greater efficiency.

Depthing tool: A tool which will accommodate two wheels or a wheel and pinion between their centers and, by means of a screw, bring them into correct pitch; this distance may then be transferred to the plates for comparison or verification.

Detent: The setting lever. Also that part of the chronometer escapement that locks the escape wheel. A detainer or pawl.

Dial train: The train of wheels under the dial which motivates the hands. The cannon pinion, hour wheel, minute wheel and pinion.

Digital display: Any timepiece where time is read with digits, for example, 12:52 as on lighted public time displays made up of light bulbs. On watches this may be by a rotating disc, light-emitting diodes or liquid crystals.

Diamantine: A powdered, crystallized boron used to give the dead flat black polish to steel.

Diode: An electronic valve used in electric watches mostly to suppress the spark when contact is broken.

Discharging pallet: The exit pallet jewel. The pallet jewel from which an escape tooth drops as it leaves the pallet.

Dividing plate: A multi-holed round plate used in wheel cutting to index the cutting of wheel teeth so that the number of teeth to be cut are equally divided. Also called an index plate.

Dog: A yoke with a set screw used to drive work held between centers in the lathe.

Dovetail: A type of slot cut into the space previously occupied by a broken wheel tooth.

Double roller: Two discs mounted on the balance staff, the smaller, crescented disc set above the larger which contains the impulse roller jewel.

Douzieme: A twelfth part of a ligne, 1/144th part of a Paris inch. A Swiss unit used to gauge movement thicknesses.

Draw: The force which keeps the pallet against the banking pins. The result of the combined angles of the escape teeth and the pallet locking surface.

Drop: The free, unrestrained motion of the escape wheel as it leaves one pallet jewel before it drops upon the locking surface of another pallet jewel.

Drop Lock: The distance up from the entrance corner on the locking surface of the pallet jewel where the escape wheel tooth first contacts (after drop).

Duofix: A type of hole jewel and cap jewel combination in which the cap jewel is held in place by a lock spring. Mistaken for shock proofing system.

Duplex escapement: A frictional escapement with sunburst style escape wheel. The escape wheel gives direct impulse to the balance in alternate vibrations.

Dynamic Poise: Poising a balance by observing its errors on a timing machine while the watch is running; poising while balance is in motion.

Ebauche: A term used by Swiss manufacturers to denote the raw movement without jewels, escapement, plating, engraving. The ebauche manufacturers, such as *A. S., Fontainemelon, Kurth, Fleurier,* supply their ebauches to trade name importers in the U.S.A. and other countries who have them finished, jeweled, dialed, cased, etc., and engraved with their own (advertised) name brands.

Electric watch: A wristwatch which uses the electromagnetic principle to impart impulse to a (motor) balance.

Electronic watch: A watch which uses electronic valves as contactless switches to impart movement and without which (transistor diode, etc.) the watch will not operate.

Elinvar: An alloy of nickel, steel, chromium, manganese and tungsten used in balances and springs, capable of close temperature adjustments.

Endshake: The free up and down space of pivoted wheels or arbors in their bearings.

Entrance jewel: The jewel first contacted by an escape tooth before it enters between the pallets. Also called the *right jewel.* Also called *stone.*

Epicycloid: A curve generated by a point on the circumference of a circle rolling upon another circle. The curves of train wheel teeth tops.

Equidistant escapement: The lever escapement whose pallet jewels have their entrance corners equally distant from the center of the pallet. Used more often in Swiss watches.

Face plate: A large plate mounted on a lathe chuck with three separate adjustable clamps which may be arranged to hold a watch plate or bridge for milling, drilling, jeweling, etc.

Facing: A lathe operation where a cut is taken across the face of object held in the lathe.

Fork: The upper end of the pallet containing the slot, horns and guard finger.

Fourth wheel: Usually the wheel upon which is mounted the second hand.

Frazing: A hobbing operation performed in a machine to reduce the diameter of a toothed wheel.

Guard pin: A thin finger emerging from a boss below the slot in the pallet fork and working in conjunction with the safety roller to aid in preventing "overbanking."

Hairspring: The spiraled spring attached to the balance to govern the speed of the balance oscillations.

Hanging barrel: *See* Floating barrel.

Headstock: The main part of the lathe resting on the bed and carrying the spindle and chuck.

Helical hairspring: The spiraled cylindrical spring used in marine chronometer balances.

Horns: The curved tips of the pallet fork.

Horology: The science and study of time measurement.

Hour wheel: A flat, brass, toothed wheel mounted on a tube which fits over the cannon pinion and supports the hour hand.

Hunting: A type of watch, usually a pocket watch case with front cover, which can be released by depressing the crown and with stem-winding at the figure 3 position.

Impulse pin: The roller jewel.

Incabloc: Trade name for a shock-resisting arrangement of balance jewels and staff design.

Index: The regulator.

India stone: A fast-cutting artificial abrasive.

Invar: An alloy similar to Elinvar.

Involute: A curve traced by the end of a string as it is unwound from a spool. The shape of wheel teeth used in gears where great strength is needed.

Isochronism: Quality of keeping equal time during the normal run of the mainspring, usually the qualities of a well-formed overcoil hairspring.

Jacot tool: A small hand lathe or turns used to shape pivots by burnishing.

Jasper stone: A red-brown quartz containing iron oxide capable of polishing steel, gold and other metals to a high finish.

Key, bench: A tool with varied size prongs capable of fitting into all sizes of winding arbors (negative setting).

Lantern pinion: A pinion made of pins set upright between two discs and used in cheaper clocks.

Lever: Usually referred to as the pallet.

Lift: The motion given to the pallet during contact with the tops of the escape wheel.

Light-emitting-diode: A digital time display made up of lighted bars or dots.

Line of centers: A line drawn through the escape wheel, pallet and balance centers.

Liquid crystals: A digital display made up of bars or segments which reflect light with an electrical charge. Some liquid crystal displays allow light to pass through these segments when an electrical charge is activated.

Locking spring: A spring used to lock the setting lever while it is in the setting position.

Main wheel: An ambiguous term applied occasionally to the crown wheel, the center wheel, and sometimes to the barrel.

Maltese cross: The cross-like part of the stopworks attached to the barrels of fine watches. (Geneva stopworks).

Marine chronometer: A boxed watch clock set in gimbals utilizing the spring detent escapement; used on shipboard to determine longitude.

Meantime screws: The adjustable screws in a better grade balance used to bring the watch to close time without the use of the regulator. Sometimes called *timing screws.*

Micron: One thousandth of a millimeter (0.001mm). Generally used to describe the thickness of electroplating; most often as 20 microns gold. In watchmaker's terms equal to 0.02mm. A millimeter measures 0.03937 inches.

Milling machine: A machine used to cut gears and perform milling or contouring operations.

Minute repeater: A striking watch that will ring the time to the minute by a series of gongs activated by a plunger or push piece. A watch striking the hours, quarter hours, and additional minutes.

Minute wheel: The wheel in the dial train that connects the cannon pinion with the hour wheel.

Movement rest: A platform or vise upon which the movement is placed while it is being repaired.

Negative setting: The setting mechanism that shifts into the setting position when the movement is out of the case. Employing a hollow winding arbor, the setting mechanism is locked into the winding position only when the movement is in the case and the case sleeve and stem are properly set.

Nonmagnetic: A balance and spring composed of alloys that will not retain magnetism after being put through a magnetic field.

Oilstone: Generally, the Arkansas white stone used with oil.

Open face: An uncovered watch dial with the figure "12" at the winding stem.

Oscillating weight: The swinging weight on self winding watches. Generally referred to those which swing between bumper springs and not a full 360°.

Overbanking: The malfunction of the pallet fork in which it shifts from one banking pin to another without being released by the roller jewel.

Overcoil: The Breguet type of hairspring.

Pallet: The jeweled lever working in conjunction with the escape wheel; the frame containing the pallet jewels.

Pedometer: A watch with a delicately tripped pendulum and ratchet attached used to tell walking distances covered by the wearer. The jogging motion moves the pendulum which advances a set of gears attached to an indicator. Later this principle was used to wind a watch called the "self-winding watch."

Pegwood: Small wood stick about ⅛" thick and 6" long whose sharpened end is used to clean or "peg" jewel holes.

Pillar plate: The lower or main plate of the watch.

Pin pallet: The lever escapement wherein the pallet has upright pins instead of horizontally set jewels. Used in alarm clocks and nonjeweled cheaper watches.

Pinion: The small geared arbor of a wheel pushed by a larger wheel.

Pivot: The thin end of a moving axle or arbor.

Poising: An operation to adjust the balance so that all weights are counterpoised.

Position timing: Adjusting a watch so that it keeps precise time when the watch is placed in a given position.

Potence: The little shelf used to support the lower balance pivot and jewels in 18s watches.

Quartz Crystal Clock: A crystal controlled synchronous clock of extremely high accuracy. "Atom" clocks are now more accurate, based upon vibrations of atoms of cesium.

Quartz timepiece: An electronic watch or clock whose accuracy is controlled by the piezo-electric effect of an especially ground quartz crystal.

Ratchet: A wheel usually placed over a mainspring arbor and working with a retaining click or pawl.

Receiving pallet: The entrance pallet jewel.

Recoil click: A click designed so that it will not permit the mainspring to be wound dead tight, recoiling a bit after any winding.

Regulator: Part of the balance bridge which resembles a racquette (racket) and contains vertical pins which straddle the hairspring. When the regulator moves towards the stud, the effective length of the hairspring is made longer and the balance slows in speed; when the pins are moved farther from the stud, the hairspring is made shorter and the watch goes faster.

Repeater: A watch that will ring the time when a push piece is motivated.

Rose cutter: A hollow, steel tube with cutting teeth around its top.

Roskopf watch: A watch with the barrel encroaching upon the center of the movement; hence no center wheel. The dial train is activated through the minute wheel which is set clutch tight on the barrel cover. Usually employing the pin pallet escapement.

Rotor: The swinging weight in self-winding watches which turns fully in a complete arc of 360°, as differentiated from the term, "oscillating weight" whose arcs are restricted by bumpers.

Ruby pin: The upright roller jewel set into the impulse roller.

Run: A term applied to the action of *slide* caused by *draw*. The action of the pallet toward the banking pin after *lock* takes place.

Safety roller: The small, crescented roller disc planted above the impulse roller. The upper part of the double roller.

Screwhead tool: A polishing tool used to finish the tops of screwheads.

Screwplate: A steel plate with holes of many sizes threaded with cutting edges for the forming of watch screws.

Setting lever: The detent which fits into the slot of the stem and pushes down the clutch lever.

Shake: The distance the escape wheel can be moved backward (manually) before the back of a tooth contacts a pallet jewel when the opposite jewel is at the very moment of unlocking.

Sidereal time: The standard of time used by astronomers. A sidereal day is 3 minutes, 56.9 seconds faster than a mean solar day.

Single roller: A roller containing both the jewel and the safety crescent on one disc. Discarded recently in favor of the double roller.

Slide: *See* Run.

Slide rest: A semi-automatic tool attached to the lathe bed and holding a cutter which may be advanced by turning a handle attached to a threaded post or lead-screw. A compound rest.

Snailing: Decorative abrading of wheel and plates in pattern form.

Solar time: The time recorded by shadows cast by gnomon or sundials. *Mean* solar time is the time used in everyday life; an average or equal measure of time.

Sonic watch: A watch or clock using a tuning fork or reed as its frequency stand-ard and usually aiding in indexing the train of wheels.

Spirit level: A small sealed disc with a liquid bubble used on poising tools to determine when the jaws are level.

Split chuck: The modern lathe collet or chuck partially "split" or cut in three equally spaced parts through the front of the chuck to give it resiliency and some tolerance.

Spotting: To decorate a plate or wheel by a series of equally spaced spots or whirls made by a revolving abrading rod.

Springing: The act of co-ordinating a hairspring with a balance so that the vibra-tions will equal a given number per hour; also called *vibrating*.

Staff: A pivoted arbor or axle usually referred to the axle of the balance; as the "balance staff."

Stake: A metal die holder or anvil or rest.

Star wheel: A steel wheel used in chronographs to lift levers.

Steady pins: The pins on the bottom of bridges used to give stability to the bridge and to prevent its shifting.

Stem: The squared shaft going through the winding pinion and clutch wheel.

Step chuck: A lathe chuck with steps of graduating diameters used to hold wheels or other flat, round objects.

Stopwatch: A simple form of chronograph with controlled starting and stopping of the hands; sometimes also stopping the balance wheel.

Stopwork: The mechanism on a barrel of a watch or clock that permits only the central portion of the mainspring to be wound thus utilizing that portion of the spring whose power is less erratic. (Maltese cross Geneva stopworks.)

Stripping: A lathe operation that removes a very slight amount of metal from a turning surface. Usually performed with a highly polished graver on jewel set-tings to give a very bright cut.

Stud: The metal piece anchored to the balance bridge into which the outer end of the hairspring is attached.

Sunk seconds: The small second dial which is depressed to avoid the second hand from interfering with the progress of the hour and minute hand.

Tailstock: The body on the lathe bed opposite the headstock used to support work held between centers or long rods. Also may contain drills or other attachments.

Tempering: Preparing metal for a desired hardness by heating or hammering.

Tetrachlorethylene: (CCl_2: CCl_2) A hydrocarbon used in watch cleaning. Also known as *Perk* and Perchlorethylene. Non inflammable and less toxic than carbon tetrachloride. Much used as a rinse.

Third wheel: The train wheel between the center and fourth wheel.

Timing screws: Adjustable balance screws used to decrease or increase the effective diameter of the balance in order to retard or hasten its vibrations.

Total lock: The distance of lock upon the pallet jewel after *slide* when the pallet rests against the banking pin.

Tourbillion: A watch in which the escapement, mounted on a cage attached to the fourth pinion, revolves around the mounted and stationary fourth wheel.

Train: A set of wheels geared or connected together.

Transducer: That part of such a device that receives energy from the electronic generator to create the sound vibrations. *See* Ultrasonics.

Tripoli: A decomposed siliceous limestone used as a polishing powder; also known as *rottenstone*.

Tripping: A malfunction caused by the failure of the escape tooth to lock upon the locking surface of the pallet jewel. Instead, the tooth enters directly upon the lifting surface with the result that the pallet may have an action like an alarm clock hammer.

Tuning fork watch: See "Sonic watch."

Turns (Turning): A lathe, generally referred to the old English bow operated lathe with dead centers and ferrule.

Ultrasonics: Cleaning watch parts by very high frequency sound waves (over 18,000 vibrations a second) traveling through a liquid in which the parts are immersed. Frequencies from 22,000 (22Kc) to one million (1 meg.) are used.

Up-and-down indicator: The semi-circular dial on chronometers that tells how much the mainspring has been unwound and thus indicates when the spring should be wound. Also called Reserve power indicator in automatic watches.

Verge: An outmoded recoil frictional escapement with a crown escape wheel and pallets set at right angles to the axis of the escape wheel.

Vibrating tool: A master balance of certified accuracy as to vibrations per hour which is mounted in a box with glass top. The box may be swiveled to set the balance into its vibratory arcs. The balance to be compared or vibrated is suspended by its hairspring attached to a scaffold and when the box is twisted on its platform both balances will start vibrating. Thus the suspended balance may be compared (in speed) with the master balance and its hairspring lengthened or shortened until both balances swing in unison.

Wax chuck: A cement chuck on which objects that cannot be held accurately in the split chuck are cemented to a brass rod held in this chuck.

Wig wag: A small machine or tool to polish pinion leaves by a back-and-forth motion. Also called *polishing shovels* when used in conjunction with an eccentric pin set into the face of a tailstock taper.

Winding pinion: The first winding wheel through which the stem enters. A wheel with two sets of teeth. One is set radial to its center and the other is set upright, crown style with ratchet teeth. The wheel above the clutch wheel.

Yoke: Part of the setting mechanism which holds down other setting parts. Also called the *setting bridge* and sometimes (incorrectly) meaning the clutch lever.

II: Watch Sizes

A watch movement is measured by gauging it across the dial side of the main (lower) plate through its center at its narrowest width. In present use, there are two different units of measure applied to the size of a watch movement. The Swiss use the ligne as their unit. The English and American manufacturers base theirs upon the full diameter of a round "O" size movement.

The Swiss ligne is one-twelfth of a Paris inch. The 0 size American movement gauges $1\frac{5}{30}$ of an English inch. Every progressive size will be $\frac{1}{30}''$ greater in diameter. Thus a 1-size movement would measure $1\frac{6}{30}''$. A movement one size smaller than the 0 size would gauge $\frac{1}{30}$ less than the standard and would be called a 2/0 size, which would then measure $1\frac{5}{30} - \frac{1}{30} = 1\frac{4}{30}$. All watches progressively smaller have zeros added to their size such as 3/0, 4/0, etc.

$$\frac{1}{30}'' = 0.84664 \text{ mm.} = 0.3753166 \text{ French lignes}$$
$$\text{One French ligne} = 2.25583 \text{ mm. or } 0.088814''$$

AMERICAN WATCH SIZES

Watch size	Fractions of an inch	Decimal inches	Millimeters	French lignes (approx.)
36/0	0	0.0	0.00	0
35/0	1/30	0.0333	0.84664	2/5
34/0	2/30	0.0666	1.69328	3/4
33/0	3/30	0.1000	2.5399	1¼
32/0	4/30	0.1333	3.3866	1½
31/0	5/30	0.1666	4.2332	1⅞
30/0	6/30	0.2000	5.0799	2¼
29/0	7/30	0.2333	5.9262	2⅔
28/0	8/30	0.2666	6.7728	3
27/0	9/30	0.3000	7.6194	3⅓
26/0	10/30	0.3333	8.4664	3¾
25/0	11/30	0.3666	9.3131	4⅛
24/0	12/30	0.4000	10.1597	4½
23/0	13/30	0.4333	11.0063	4⁹⁄₁₀
22/0	14/30	0.4666	11.8529	5¼
21/0	15/30	0.5000	12.6996	5⅔
20/0	16/30	0.5333	13.5462	6

AMERICAN WATCH SIZES

Watch size	Fractions of an inch	Decimal inches	Millimeters	French lignes (approx.)
19/0	17/30	0.5666	14.3928	6⅖
18/0	18/30	0.6000	15.2395	6¾
17/0	19/30	0.6333	16.0862	7⅛
16/0	20/30	0.6666	16.9329	7½
15/0	21/30	0.7000	17.7795	7⁹⁄₁₀
14/0	22/30	0.7333	18.6259	8¼
13/0	23/30	0.7666	19.4727	8⅔
12/0	24/30	0.8000	20.3193	9
11/0	25/30	0.8333	21.1670	9⅖
10/0	26/30	0.8666	22.0136	9¾
9/0	27/30	0.9000	22.8603	10⅛
8/0	28/30	0.9333	23.7069	10½
7/0	29/30	0.9666	24.5536	10⁹⁄₁₀
6/0	1-	1.0000	25.3999	11¼
5/0	1-1/30	1.0333	26.2461	11⅔
4/0	1-2/30	1.0666	27.0928	12
3/0	1-3/30	1.1000	27.9394	12⅖
2/0	1-4/30	1.1333	28.7861	12¾
0	1-5/30	1.1666	29.6327	13⅐
1	1-6/30	1.2000	30.4793	13½
2	1-7/30	1.2333	31.3260	13⁹⁄₁₀
3	1-8/30	1.2666	32.1726	14¼
4	1-9/30	1.3000	33.0193	14⅔
5	1-10/30	1.3333	33.8660	15
6	1-11/30	1.3666	34.7125	15⅖
7	1-12/30	1.4000	35.5592	15¾
8	1-13/30	1.4333	36.4058	16⅛
9	1-14/30	1.4666	37.2526	16½
10	1-15/30	1.5000	38.0993	16⁹⁄₁₀
11	1-16/30	1.5333	38.9459	17¼
12	1-17/30	1.5666	39.7925	17⅔
13	1-18/30	1.6000	40.6392	18
14	1-19/30	1.6333	41.4858	18⅖
15	1-20/30	1.6666	42.3325	18¾
16	1-21/30	1.7000	43.1791	19⅐
17	1-22/30	1.7333	44.0257	19½
18	1-23/30	1.7666	44.8724	19⁹⁄₁₀
19	1-24/30	1.8000	45.7190	20¼
20	1-25/30	1.8333	46.5652	20⅔
21	1-26/30	1.8666	47.4118	21
22	1-27/30	1.9000	48.2585	22⅖
23	1-28/30	1.9333	49.1051	22¾
24	1-29/30	1.9666	49.9518	23⅐
25	2-	2.0000	50.7999	23½

SWISS WATCH SIZES

Lignes	Size (American approx.)	Millimeters	Inches (decimal)
¼	35/0	0.5639775	0.0222035
½		1.1279	0.0444
¾	34/0	1.6919	0.0666
1		2.2558	0.0888
1¼	33/0	2.8198	0.1111
1½	32/0	3.3384	0.1332
1¾	31/0	3.9477	0.1554
2		4.5117	0.1776
2¼	30/0	5.0757	0.1998
2½		5.6396	0.2220
2¾	29/0	6.2036	0.2442
3	28/0	6.7675	0.2664
3¼	27/0	7.3315	0.2886
3½		7.8954	0.3108
3¾	26/0	8.4594	0.3330
4		9.0233	0.3553
4¼	25/0	9.5873	0.3775
4½	24/0	10.1512	0.3997
4¾		10.7152	0.4219
5	23/0	11.2792	0.4441
5¼	22/0	11.8432	0.4663
5½		12.4071	0.4885
5¾	21/0	12.9708	0.5107
6	20/0	13.5350	0.5329
6¼		14.0990	0.5551
6½	19/0	14.6629	0.5773
6¾	18/0	15.2266	0.5995
7		15.7908	0.6217
7¼	17/0	16.3548	0.6439
7½	16/0	16.9187	0.6661
7¾		17.4827	0.6883
8	15/0	18.0466	0.7105
8¼	14/0	18.6108	0.7327
8½		19.1745	0.7549
8¾	13/0	19.7385	0.7771
9	12/0	20.3025	0.7993
9¼		20.8665	0.8215
9½	11/0	21.4304	0.843?
9¾	10/0	21.9944	0.8659
10	9/0	22.5583	0.8881
10¼		23.1223	0.9103
10½	8/0	23.6862	0.9325
10¾		24.2502	0.9547
11	7/0	24.8141	0.9770
11¼	6/0	25.3781	0.9992

SWISS WATCH SIZES

Lignes	Size (American approx.)	Millimeters	Inches (decimal)
11½		25.9420	1.0214
11¾	5/0	26.5060	1.0436
12	4/0	27.0700	1.0658
12¼		27.6340	1.0880
12½	3/0	28.1979	1.1102
12¾	2/0	28.7619	1.1324
13	0	29.3258	1.1546
13¼		29.8998	1.1768
13½	1	30.4537	1.1990
13¾		31.0177	1.2212
14	2	31.5816	1.2434
14¼	3	32.0456	1.2656
14½		32.7095	1.2878
14¾	4	33.2732	1.3100
15	5	33.8375	1.3322
15¼		34.4015	1.3544
15½	6	34.9654	1.3766
15¾	7	35.5291	1.3988
16		36.0933	1.4210
16¼	8	36.6573	1.4432
16½	9	37.2212	1.4654
16¾		37.7852	1.4876
17	10	38.3491	1.5098
17¼	11	38.9131	1.5320
17½		39.4770	1.5542
17¾	12	40.0410	1.5764
18	13	40.6049	1.5987
18¼		41.1689	1.6209
18½	14	41.7328	1.6431
18¾	15	42.2968	1.6653
19		42.8608	1.6875
19¼	16	43.4247	1.7097
19½	17	43.9887	1.7319
19¾		44.5527	1.7541
20	18	45.1166	1.7763
21	21	47.3724	1.8651
22	23	49.6283	1.9539
23	26	51.8841	2.0427
24	29	54.1399	2.1315
25	32	56.3958	2.2204
26	34	58.6516	2.3092
27	37	60.9074	2.3980
28	40	63.1632	2.4868
29	43	65.4191	2.5756
30	45	67.6749	2.6644

CONVERSION TABLES

Millimeters to Inches to Lignes
Inches to Millimeters to Lignes
Lignes to Inches to Millimeters

MILLIMETERS TO INCHES TO LIGNES

Millimeters	Inches	Lignes	Millimeters	Inches	Lignes
0.01	0.0003937	0.004433	17.00	0.66929	7.5361
0.02	0.0007874	0.008866	18.00	0.70866	7.9794
0.03	0.0011811	0.013299	19.00	0.74803	8.4227
0.04	0.0015748	0.017732	20.00	0.78740	8.8660
0.05	0.0019685	0.022165	21.00	0.82677	9.3093
0.06	0.0023622	0.026598	22.00	0.86614	9.7526
0.07	0.0026559	0.031031	23.00	0.90551	10.1959
0.08	0.0031496	0.035464	24.00	0.94488	10.6392
0.09	0.0035433	0.039897	25.00	0.98425	11.0825
0.10	0.003937	0.04433	26.00	1.02362	11.5258
0.20	0.007874	0.08866	27.00	1.06299	11.9691
0.30	0.011811	0.13299	28.00	1.10236	12.4124
0.40	0.015748	0.17732	29.00	1.14173	12.8557
0.50	0.019685	0.22165	30.00	1.18110	13.2990
0.60	0.023622	0.26598	31.00	1.22047	13.7423
0.70	0.026559	0.31031	32.00	1.25984	14.1856
0.80	0.031496	0.35464	33.00	1.29921	14.6289
0.90	0.035433	0.39897	34.00	1.33858	15.0722
1.00	0.03937	0.4433	35.00	1.37795	15.5155
2.00	0.07874	0.8866	36.00	1.41732	15.9588
3.00	0.11811	1.3299	37.00	1.45669	16.4021
4.00	0.15748	1.7732	38.00	1.49606	16.8454
5.00	0.19685	2.2165	39.00	1.53543	17.2887
6.00	0.23622	2.6598	40.00	1.57480	17.7320
7.00	0.26559	3.1031	41.00	1.61417	18.1753
8.00	0.31496	3.5464	42.00	1.65354	18.6186
9.00	0.35433	3.9897	43.00	1.69291	19.0619
10.00	0.3937	4.4330	44.00	1.73228	19.5052
11.00	0.43307	4.8763	45.00	1.77165	19.9485
12.00	0.47244	5.3196	46.00	1.81102	20.3918
13.00	0.51181	5.7629	47.00	1.85039	20.8351
14.00	0.55118	6.2062	48.00	1.88976	21.2784
15.00	0.59056	6.6495	49.00	1.92913	21.7217
16.00	0.62992	7.0928	50.00	1.96850	22.1650

CONVERSION TABLES

Millimeters to Inches to Lignes
Inches to Millimeters to Lignes
Lignes to Inches to Millimeters

INCHES TO MILLIMETERS TO LIGNES

Inches	Millimeters	Lignes	Inches	Millimeters	Lignes
0.001	0.025399	0.011260	0.20	5.0798	2.25190
0.002	0.050798	0.022159	0.30	7.6197	3.37785
0.003	0.076197	0.033779	0.40	10.1596	4.50380
0.004	0.101596	0.045038	0.50	12.6995	5.62975
0.005	0.126995	0.056298	0.60	15.2394	6.75570
0.006	0.152394	0.067557	0.70	17.7793	7.88165
0.007	0.177793	0.078817	0.80	20.3192	9.00760
0.008	0.203192	0.090076	0.90	22.8591	10.13355
0.009	0.228591	0.101336	1.00	25.3999	11.2595
0.01	0.25399	0.112595	1.10	27.9389	12.38545
0.02	0.50798	0.225190	1.20	30.4788	13.51140
0.03	0.76197	0.337785	1.30	33.0187	14.63735
0.04	1.01596	0.450380	1.40	35.5586	15.76330
0.05	1.26995	0.562975	1.50	38.0985	16.88925
0.06	1.52394	0.675570	1.60	40.6384	18.01520
0.07	1.77793	0.788165	1.70	43.1783	19.14115
0.08	2.03192	0.900760	1.80	45.7182	20.26710
0.09	2.28591	1.013355	1.90	48.2581	21.39305
0.10	2.5399	1.12595	2.00	50.7999	22.51900

CONVERSION TABLES
Millimeters to Inches to Lignes
Inches to Millimeters to Lignes
Lignes to Inches to Millimeters

LIGNES TO INCHES TO MILLIMETERS

A ligne is equal to one-twelfth of a Paris inch. The Paris inch = 1.06577″ and 27.06996 mm. The douzième is equal to 1/144th part of a Paris inch or one-twelfth of a ligne. It is used to gauge movement thicknesses (Swiss). A douzième equals .007412″ and .187985 mm.

Lignes	Inches	Millimeters	Lignes	Inches	Millimeters
0.01	0.000888	0.0225583	2.0	0.177628	4.51166
0.02	0.001776	0.0451166	3.0	0.266442	6.76749
0.03	0.002664	0.0676749	4.0	0.355256	9.02332
0.04	0.003552	0.0902332	5.0	0.444070	11.27915
0.05	0.004440	0.1127915	6.0	0.532884	13.53498
0.06	0.005328	0.1353498	7.0	0.621698	15.79081
0.07	0.006217	0.1579081	8.0	0.710512	18.04664
0.08	0.007105	0.1804664	9.0	0.799326	20.30247
0.09	0.007993	0.2030247	10.0	0.88814	22.55830
0.10	0.008881	0.225583	11.0	0.97696	24.81413
0.20	0.017763	0.451166	12.0	1.06577	27.06996
0.30	0.026644	0.676749	13.0	1.15458	29.32579
0.40	0.035526	0.902332	14.0	1.24340	31.58162
0.50	0.044407	1.127915	15.0	1.33221	33.83745
0.60	0.053288	1.353498	16.0	1.42103	36.09328
0.70	0.062169	1.579081	17.0	1.50984	38.34911
0.80	0.071051	1.804664	18.0	1.59865	40.60494
0.90	0.079933	2.030247	19.0	1.68747	42.86077
1.0	0.088814	2.25583	20.0	1.77628	45.11660

TABLE OF COMPARATIVE METRIC AND DENNISON MAINSPRING SIZES
THICKNESS (OR "STRENGTH") OF MAINSPRINGS

Metric (mm)	Dennison No.	Metric (mm)	Dennison No.	Metric (mm)	Dennison No.
0.01	20	0.11	10	0.21	3
0.02	19	0.12	9½	0.22	2
0.03	18	0.13	9	0.23	1
0.04	17	0.14	8	0.24	0½
0.05	16	0.15	7	0.25	0
0.06	15	0.16	6½	0.26	00
0.07	14	0.17	6	0.27	000
0.08	13	0.18	5	0.28	000½
0.09	12	0.19	4	0.29	0000
0.10	11	0.20	3½	0.30	00000(5/0)

WIDTH OF MAINSPRINGS

Metric (mm)	Dennison Width	Metric (mm)	Dennison Width	Metric (mm)	Dennison Width
0.00	10/0	1.30	4	2.60	17
0.10	9/0	1.40	5	2.70	18
0.20	8/0	1.50	6	2.80	19
0.30	7/0	1.60	7	3.00	21
0.40	6/0	1.70	8	3.20	23
0.50	5/0	1.80	9	3.40	25
0.60	4/0	1.90	10	3.70	28
0.70	3/0	2.00	11	4.00	31
0.80	2/0	2.10	12	4.50	36
0.90	0	2.20	13	5.00	41
1.00	1	2.30	14	5.25	43½
1.10	2	2.40	15	5.50	46
1.20	3	2.50	16	6.00	51

TABLE OF WATCH CROWN THREADS
(as revised June 1964)

Tap No.	Major Dia. (millimeters)	Major Dia. (inches)	Threads per inch	Threads per millimeter
0	2.26	.089	60	2.3
1	1.88	.074	72	2.8
2 (pocket size)	1.55	.060	80	3.1
2 (pinlever)	1.47	.058	84	3.3
3, 6 & 7	1.19	.047	110	4.3
4	1.29	.051	84	3.3
5	1.36	.054	84	3.3
8	1.09	.043	100	3.9
9 (Elgin)	1.03	.041	120	4.7
9 (Swiss)	.99	.039	110	4.3
10 (Amer.)	.90	.036	113	4.4
10 (Swiss)	.89	.035	110	4.3
11	.84	.033	140	5.5
12	.76	.030	140	5.5

III: How to Order Watch Parts

Today, it is possible to obtain precision replacement parts to almost all watches made within the past thirty years. However, in order to obtain parts from a jeweler's and watch material supply house so that the replacements are exact, certain information must be submitted with your order. There are literally thousands of different makes and calibres of watch movements. Even the slightest change in the placement of an additional jewel such as the escape wheel cap jewel is enough to make the part unique in measurement and characteristic. Therefore to insure receiving the correct replacement the following steps should be undertaken.

1—State the part desired. For ordinary watch parts such as winding or train wheels, it may be sufficient merely to name the part. However, for complicated watches such as calendars, chronographs, alarm and self-winding watches, it is best to give the exact name of the part as well as the official reference number. The reference numbers will be found at the part indicated in the various exploded views of the various types of watches. These reference numbers are the official numbers of almost every Swiss watch as well as those made elsewhere.

2—State the maker's name and the calibre reference number. For example, should it be desired to order a part for the movement pictured in Figure 1 on the page, "How to Identify a Movement," the order should include the statement "Eta, Calibre 1256." If an oscillating weight is needed for this model, the order should read, "One part, No. 1143, Eta, Cal. 1256."

3—When ordering staffs, always specify whether it is for a watch using a shock-proofing device; the type of device employed as staffs for one make of watch may vary as much as 1/10th m/m, depending on the type of shockproofing used. When ordering escape wheels or pallet arbors, state if the pivots have cap jewels or which pivot has the (only) cap jewel. Rollers should be ordered with the specification whether it is used on a particular type of shockproofing or not. Likewise, cannon pinions or your wheels should have their heights included with the order in the case of curved dials. Specify whether the train wheels are for a watch with a sweep second hand. Center, third and fourth wheels are different from those where the second hand is not in the center of the dial.

4—When ordering mainsprings, specify whether you want an alloy (unbreakable) mainspring. If for an automatic watch, specify whether you want the spring with the bridle (slipping attachment) combined.

IV: How to Identify a Movement

Look for the manufacturer's symbol, trade mark and calibre or reference number. Shown here are some examples where these might be found.

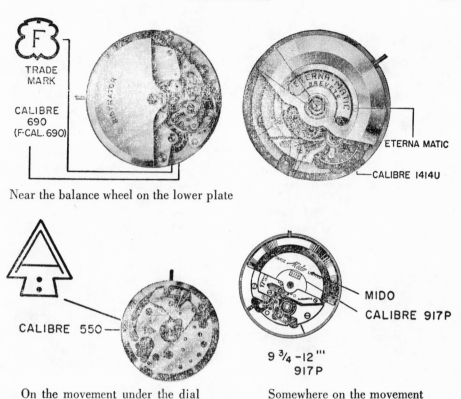

Near the balance wheel on the lower plate

CALIBRE 550 —

On the movement under the dial

$9\,\frac{3}{4}-12\,'''$
917 P

Somewhere on the movement

II: Types of Movements

Here are the 12 types of movements that watchmakers most commonly are called upon to service.

Various highly specialized timepieces—such as marine chronometers, repeater movements, multiple split-second timers, specialized aviation and navigational computers and "custom-built" models—are usually serviced by the manufacturers or a specialist.

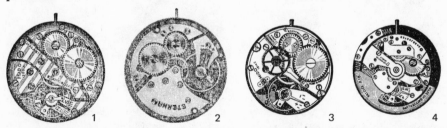

1. REGULAR MOVEMENTS—No special features; 2. DIRECT SWEEP SECOND—Fourth wheel through center wheel; 3. INDIRECT SWEEP SECOND—Additional fourth wheel on top plate driving through second pinion through hollow center pinion; 4. AUTOMATIC WATCH—Has self winding mechanism.

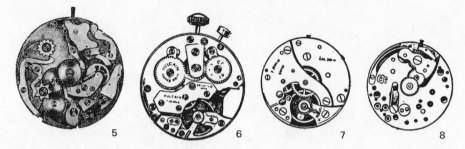

5. CALENDAR—Any watch with date, month, moon phase, etc.; 6. ALARM WATCH—Any watch or jeweled clock with alarm mechanism; 7. PIN LEVER—Any watch with vertical pins instead of pallet jewels (Roskopf Escp.); 8. BASCULE SETTING—A movement with set wheels under one plate and pivoted on them with all winding on dial side.

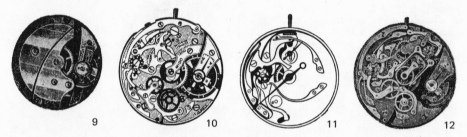

9. CYLINDER ESCAPEMENT MOVEMENT—Without any lever but with raised beak teeth enmeshing into hollow cylindrical balance staff; 10. CHRONOGRAPH—Watch with controlled sweep second hand also with separate minute or hour counters, elapsed time, etc; 11. STOP WATCH—Watch with only sweep second hand and minute counter with start, stop and fly-back control. Watch stops when hand is stopped; 12. SPLIT SECOND CHRONOGRAPH—Watch with hour, minute and second hand also with two or more sweep second hands and subcounters.

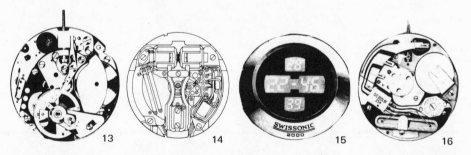

13. BATTERY-DRIVEN ELECTRIC WATCH with transistorized balance and coil; electronic switch without contacts. 14. ACCUTRON TUNING FORK WATCH with transistorized circuit. 15. QUARTZ WATCH without balance or tuning fork, by Girard Perregaux; train of wheels is indexed by a tiny stepping motor. 16. SOLID STATE, integrated circuit watch with no moving parts; digital display is by liquid crystal readout.

BIBLIOGRAPHY AND REFERENCE MATERIAL

A complete listing of horological literature would require a volume in itself. Books on horology go back to the seventeenth century, and there are many volumes with earlier dates that have references to the horological arts. However, listed here are books in the English language which are useful and considered to have contributed to the art and science of watchmaking. The fact that many of these books are out of print should not disqualify their being listed here because, although they are not readily available for purchase, they still are on the shelves of the larger public libraries and horological societies.

Horological books generally fall into definite categories. Most deal with the practical application and manual arts of repairing or construction. Others dwell upon the scientific and theoretical descriptions of the mechanisms. Other books are purely historical in content, giving the chronological advance of horology through the centuries. Still others are published to aid the collector and hobbyist to date and identify the timepieces in their possession, and some are philosophical, giving the author's concept of time. Except where the title may be confusing, no description other than the title will be given.

American Clocks and Clockmakers (Historical, antiques, lists of makers), C. W. Dreppard, Doubleday, N. Y., 1947.

American Watchmaker and Jeweler, H. C. Abbot, Hazlitt and Walker, Chicago, 1908.

Antique Watches and How to Establish Their Age, H. C. Abbot, Barny Clockmakers, 218 E. 59th St., N. Y., 1910.

Automatic Watch, The, R. W. Pipe, Heywood & Co., Ltd. 1952.

Bench Practices for Watch Repairers, Henry B. Fried, Roberts Pub. Co., Denver, 1954.

Book of American Clocks, Brooks Palmer, Macmillan & Co., 1950. Most authentic on early American clocks and watches and dates.

Calendar Watches, Modern, B. Humbert, Journal Suisse D'Horlogerie, 1954.

Cavalcade of Time, The, Henry B. Fried, Zale Corp., Dallas, Texas, 1968.

Chronograph, The, B. Humbert, Journal Suisse, Lausanne, Switzerland, 1952.

Clock Book, The, (Historical and descriptive), W. Nutting, Garden City Pub. Co., N.Y., 1935.

Clock Repairing and Making, F. J. Garrard, Crosley Lockwood & Sons, Technical Press, 1946.

Clock Repairing, Practical, Donald deCarle, N. A. G. Press Ltd. 1952.

Clock and Watch Escapements, W. J. Gazeley, D. Van Nostrand Co., Inc., Princeton, 1956.

Complicated Watches, E. Seibel, Roberts Pub. Co., Denver, 1945.

Complicated Watches and Their Repair, deCarle, N. A. G. Press Ltd., 1956.

Complicated Watches, A Guide to, F. Lecoultre, Ch. Rohr Ltd. Switzerland.

Electric Clocks, S. J. Wise, Heywood & Co., Ltd., 1951.

Electric Clocks, Modern, Stuart Philpott, Pitman, 1949.

Electric Watch Repair Manual, Henry B. Fried, B. Jadow & Sons, Inc., N.Y., 1972.

Electrical Horology, Langman & Ball, Hensley Pub. Co., N. Y., 1923

Electrical Timekeeping, Hope-Jones, N. A. G. Press, 1938.

Escapement and Train of American Watches, The, T. J. Wilkinson, Keystone Pub. Co., Phila., 1928.

Escapement, The Watch, How to Repair, Adjust, Analyse, Henry B. Fried, B. Jadow, N. Y., 1960.

Horology (Science, theory and description), E. Haswell, Chapman & Hall, Ltd., London, 1948.

Horolovar, 400 Day Clock Repair Guide, Terwilliger, Horolovar Co., Bronxville, N. Y., 1960.

It's About Time (Historical and descriptive drawings of many escapements) P. M. Chamberlain, Rich, Smith, N. Y., 1941.

Know The Escapement, S. & H. Barkus, Barkus Laboratories, Los Angeles, 1943.

Lessons in Horology, F. & H. Grossman, Keystone Pub. Co., 1905.

Lure of the Clock (Historical and description of clocks in N. Y. U. clock museum), New York University Press, N. Y., 1942.

Marine Chronometer, The, R. T. Gould, J. Potter, London, 1923.

Modern Clock, The, W. L. Goodrich, Hazlitt & Walker, Chicago, 1905.

Modern Clocks, T. R. Robinson, N. A. G. Press, Ltd., 1955.

Modern Methods in Horology, G. Hood, Bradley Univ. Press, Peoria, Ill., 1926.

Modern Watch and Clock Repairing, P. B. Harris, Nelson Hall, Chicago, 1947.

Modern Watch Repairing and Adjusting, Bowman-Borer, H. Paulson Co., Chicago, 1941.

Old Clock and Watches and Their Makers, Britten's, Baille, Clutton & Ilbert., E. & F. N. Spon Ltd., 1956.

Packard Collection of Unusual and Complicated Watches, Henry B. Fried, American Watchmakers Institute, Detroit, Mich., 1959.

Practical Balance and Hairspring Work, W. J. Kleinlein, Waltham, Mass., 1944.

Practical Benchwork for Horologists, L. & S. Levin, Levin & Son, Los Angeles, 1945.

Practical Clock Repairing, D. deCarle, N.A.G. Press, Ltd., London, 1952.

Practical Course in Horology, A, H. Kelly, Manual Arts Press, Peoria, Ill., 1944.

Practical Watch Repairing, D. deCarle, N.A.G. Press, Ltd., London, 1946.

Precision Time Measurement, Chas. Higginbotham, Hazlitt ˙& Walker, 1913.

Rules and Practices for Adjusting Watches, W. J. Kleinlein, Waltham, Mass., 1940.

Science of Clocks and Watches, The, A. L. Rawlings, Pitman Pub. Co., 1948.

Science of Watch Repairing Simplified, The, A. G. Thissel, H. Paulson Co., Chicago, 1943.

Self-Winding Watches, Humbert, Journal Suisse D'Horlogerie, Lausanne, Switzerland, 1956.

Some Outstanding Clocks Over Seven Hundred Years, 1250–1950, H. Alan Lloyd, Leonard Hill, 1958.

Swiss Watch Repairers Manual, H. Jendritzki, Journal Suisse D'Horlogerie, Lausanne, Switzerland, 1953.

Technical Manual TM 9-1575, War Dept. Manual, Ordnance-Maintenance, Wrist Watches, Pocket Watches, U. S. Government Printing Office, Wash., D. C., 1945.

Time and Its Measurement, Harrison J. Cowen, World Pub. Co., 1958.

Time Telling Through The Ages, H. C. Brearly, Doubleday, Page Co., 1920.

Time and Timekeepers and Their Makers, (Historical, descriptive, comprehensive bibliography), W. I. Milham, Macmillan Co., N. Y., 1923.

Training Manual, Jos. Bulova School of Watchmaking, Bulova W. Co., Woodside, N. Y., 1947.

Treatise on Modern Horology (an encyclopedic work), Claudius Saunier, Crosley Lockwood & Co., London, 1887.

Universal Watch Parts Catologue, H. B. Fried, Watch Material Distributors Association of America, Wash. 5, D. C., 1957.

Watch Adjuster Manual, The, C. E. Fritts, Keystone Pub. Co., Phila., 1912.

Watch and Clock Encyclopedia, D. deCarle, N. A. G. Press, Ltd., 1959.

Watch and Clockmakers Handbook, (Dictionary & guide), 15 edition, F. J. Britten, D. Van Nostrand Co., N. Y., 1955.

Watch and Clockmakers of the World, G. H. Baille, (A listing of names, dates and places of makers), N. A. G. Press, Ltd., 1948.

Watch Escapements, James C. Pellaton, N. A. G. Press, Ltd., 1950.

Watch Repair, H. C. Kelly, Chas. W. Bennet Co., 1957.

Watch Repairing, J. W. Player, Crosley Lockwood & Sons, Ltd., 1945.

Watch Repairing as a Hobby, D. W. Fletcher, Pitman Pub. Corp., 1947.

Watch Repairing, Cleaning and Adjusting, F. J. Garrard, E. and F. N. Spon, Ltd., London, 1922.

Watches, G. H. Baille, (historical) Methuen & Co., Ltd., London, 1929.

Watches, Adjustment and Repair, F. J. Camm, Chemical Pub. Co., Brooklyn, 1945.

Watches, The Country Life Book of, T. P. Camerer Cuss, Country Life Ltd., London, 1967.

Watchmakers Lathe, The, W. L. Goodrich, North American Watch Tool Co., Chicago, 1903, 1956.

Watchmakers Lathe, The, and How To Use It, D. deCarle, N. A. G. Press, Ltd., 1952.

Waterproofing Principles and Practices, Henry B. Fried, B. Jadow & Sons, Inc., N.Y., 1962.

CURRENT TRADE PERIODICALS PUBLISHED ON A NATIONAL SCALE

American Horologist and Jeweler, 2403 Champa St., Denver, Colo. 80205.
British Horological Journal, Upton, Newark, Notts, England.
Bulletin, National Assoc. of Watch and Clock Collectors, Box 33, Columbia, Pa. 17512.
Jewelers' Circular-Keystone, Chilton Way, Radnor, Pa. 19089.
Modern Jeweler, 230 East 44th St., N.Y. 10017.
National Jeweler, 222 Park Avenue South, N.Y. 10003.
The Swiss Watch Journal (English Edition), Overseas Pub. Co., 6 West 57th St., N.Y. 10017.

NATIONAL TRADE ASSOCIATIONS

American Gem Society, 3142 Wilshire Blvd., Los Angeles, Calif. 90010.
American Watch Association, 39 Broadway, N.Y. 10016.
American Watchmakers Institute, Box 11011, Cincinnati, Ohio 45211.
British Horological Institute, Upton, Newark, Notts, England.
Canadian Jewellers Association, 663 Yonge St., Rm. 401, Toronto 285, Ontario.
Gemological Institute of America, 11940 San Vincente Blvd., Los Angeles, Calif. 90049.
The Horological Guild of Australasia, 346 Lt. Collins St., Melbourne, Australia.
Jewelry Industrial Council, 608 Fifth Ave., N.Y. 10020.
National Association of Watch and Clock Collectors, Box 33, Columbia, Pa. 17512.
Watch Material and Jewelry Distributors Association, 2135 Wisconsin Ave., N.W., Washington, D.C. 20007.
Watchmakers of Switzerland Information Center, The, 608 Fifth Ave., N.Y. 10020.

INDEX

(Items not found in this index might be listed in the "Dictionary of Trade Terms" on page 290, since this index does not list the individual terms contained in the dictionary.)

HENRY B. FRIED

Certified Master Watchmaker, Certified Master Clockmaker, Henry B. Fried is a third-generation watchmaker who has studied with New York's finest craftsmen. He has held positions as head watchmaker and trade shop operator before being licensed to organize and teach the first horology class for the New York City Board of Education, a position he held for more than 30 years. He was head of the Department of Horology at the George Westinghouse Vocational and Technical High School in New York City. He is the author of many textbooks and numerous articles on timekeepers of the past, present and contemplative future. Mr. Fried illustrates many of his own texts with fine isometric drawings.

The author has written for such publications as the *Reader's Digest* as well as consumer publications and his articles have appeared in almost every trade publication in America as well as in the foreign trade press. *The Watch Repairer's Manual* has even been printed in Chinese. At present, he is the contributing editor and horological consultant-editor for the *Jewelers' Circular-Keystone*. He is also the accredited horological consultant and contributor for both the Random House and Merriam-Webster dictionaries. He is also the Technical Director of the American Watchmakers Institute and conducts the Answer Box for the National Association of Watch and Clock Collectors. Mr. Fried received the Outstanding Achievement Award from the United Horological Association of America and his numerous citations and awards from various state and national associations as well as those in Canada, Australia and Puerto Rico make him the most-recognized living horologist. In 1970, he was chosen to receive the "Man of the Year Award" of the jewelry industry (WMDJAA). His professional achievements include the past presidency of the Horological Society of New York, The New York State Watchmakers Association, vice president of the Horological Institute of America and their technical director and, currently, President Emeritus of the American Watchmakers Institute, having been their president in 1971.

A collector of all forms of horologia, Mr. Fried is also a founder member and Fellow of the National Association of Watch and Clock Collectors, as well as a Fellow of The British Horological Institute. The author also lectures extensively, has often appeared on television and radio in connection with horology. He is a graduate of the Industrial Teacher's Training College of the University of the State of New York and has also attended Queens College and Oswego University.

His other books are: *Bench Practices for Watch Repairers; The Watch Escapement; Principles and Practices of Waterproofing Watches;* Escapement and Repair section of the *400 Day Clock Repair Guide; The Universal Watch Parts Catalog; The Cavalcade of Time; The Electric Watch Repair Manual; The Packard Collection of Complicated and Unusual Watches.*